T0292128

CAMBRIDGE LIBRARY COLLECTION

Books of enduring scholarly value

Botany and Horticulture

Until the nineteenth century, the investigation of natural phenomena, plants and animals was considered either the preserve of elite scholars or a pastime for the leisured upper classes. As increasing academic rigour and systematisation was brought to the study of 'natural history', its subdisciplines were adopted into university curricula, and learned societies (such as the Royal Horticultural Society, founded in 1804) were established to support research in these areas. A related development was strong enthusiasm for exotic garden plants, which resulted in plant collecting expeditions to every corner of the globe, sometimes with tragic consequences. This series includes accounts of some of those expeditions, detailed reference works on the flora of different regions, and practical advice for amateur and professional gardeners.

Self-Instruction for Young Gardeners, Foresters, Bailiffs, Land-Stewards, and Farmers

Intended for young men with limited formal education, this manual was the final project of the landscape gardener John Claudius Loudon (1783–1843). Completed by friends, the book appeared posthumously in 1845. The son of a farmer, Loudon was well aware that men who began their careers as gardeners often became the stewards of estates, bailiffs, or tenant farmers later in life, and he provides here some of the mathematical and technical instruction necessary to carry out those roles successfully. Including sections on fractions, geometry, trigonometry, architectural drawing, and the calculation of wages and interest rates, the book traces a remarkable picture for the modern reader of the administrative duties expected of horticultural and agricultural workers in the mid-nineteenth century. Also included are conversion tables, a biography of Loudon, and a short preface by his wife Jane, whose *Instructions in Gardening for Ladies* (1840) is also reissued in this series.

Cambridge University Press has long been a pioneer in the reissuing of out-of-print titles from its own backlist, producing digital reprints of books that are still sought after by scholars and students but could not be reprinted economically using traditional technology. The Cambridge Library Collection extends this activity to a wider range of books which are still of importance to researchers and professionals, either for the source material they contain, or as landmarks in the history of their academic discipline.

Drawing from the world-renowned collections in the Cambridge University Library and other partner libraries, and guided by the advice of experts in each subject area, Cambridge University Press is using state-of-the-art scanning machines in its own Printing House to capture the content of each book selected for inclusion. The files are processed to give a consistently clear, crisp image, and the books finished to the high quality standard for which the Press is recognised around the world. The latest print-on-demand technology ensures that the books will remain available indefinitely, and that orders for single or multiple copies can quickly be supplied.

The Cambridge Library Collection brings back to life books of enduring scholarly value (including out-of-copyright works originally issued by other publishers) across a wide range of disciplines in the humanities and social sciences and in science and technology.

Self-Instruction for Young Gardeners, Foresters, Bailiffs, Land-Stewards, and Farmers

With a Memoir of the Author

<small>John Claudius Loudon</small>

<small>CAMBRIDGE UNIVERSITY PRESS</small>

CAMBRIDGE
UNIVERSITY PRESS

University Printing House, Cambridge, CB2 8BS, United Kingdom

Published in the United States of America by Cambridge University Press, New York

Cambridge University Press is part of the University of Cambridge.

It furthers the University's mission by disseminating knowledge in the pursuit of
education, learning and research at the highest international levels of excellence.

www.cambridge.org
Information on this title: www.cambridge.org/9781108066396

© in this compilation Cambridge University Press 2013

This edition first published 1845
This digitally printed version 2013

ISBN 978-1-108-06639-6 Paperback

J. C. Loudon

London, Published 1845, by Longman, Brown, Green, & Longmans.

SELF-INSTRUCTION

FOR

YOUNG GARDENERS,

FORESTERS, BAILIFFS,

LAND-STEWARDS, AND FARMERS;

IN

ARITHMETIC AND BOOK-KEEPING,
GEOMETRY, MENSURATION, AND PRACTICAL TRIGONOMETRY,
MECHANICS, HYDROSTATICS, AND HYDRAULICS,
LAND-SURVEYING, LEVELLING, PLANNING, AND MAPPING,
ARCHITECTURAL DRAWING, AND ISOMETRICAL PROJECTION
AND PERSPECTIVE:

With Examples,

SHOWING THEIR APPLICATION TO HORTICULTURAL AND
AGRICULTURAL PURPOSES.

BY THE LATE J. C. LOUDON, F.L.S. H.S. &c.

WITH A MEMOIR OF THE AUTHOR.

ILLUSTRATED WITH NUMEROUS ENGRAVINGS.

LONDON:
LONGMAN, BROWN, GREEN, AND LONGMANS,
PATERNOSTER-ROW.
1845.

London:
Printed by A. Spottiswoode,
New-Street-Square.

PREFACE.

The first idea of the present work was originated by Mr. Osborn of the Fulham Nursery, who, having for several years had numerous young men under his superintendence, perceived how much a book of the kind was wanted; and Mr. Loudon, always eager to adopt any plan that seemed likely to improve young gardeners, was so much pleased with his suggestions that he readily undertook it, making great additions and improvements in the original plan: and, had he lived a little longer, he would have made it still more complete; as he intended to include instructions in drawing landscapes, trees, flowers, &c., sketching from life, and colouring, together with the art of imitating flowers and fruits in wax, and many other similar acquirements, which are all more or less useful to a young gardener, and at the same time adapted to relieve his severer studies.

For several months after Mr. Loudon's death I could not summon courage to look at the manuscript of this work, it was so associated with the idea of his sufferings; at length, however, our kind friend Mr. Paxton, after exerting himself most warmly in our behalf, urged me to publish it; adding, what is certainly the fact, that it may be considered as my poor husband's last legacy to gardeners. I have therefore presented it to the public; and, as I really could not prepare the manuscript for the press myself, it was first looked over and arranged by Mr. Wooster (Mr. Loudon's last amanuensis), and after-

wards submitted to Dr. JAMIESON, whose name was mentioned by Mr. Loudon a few hours before his death, as the
person he wished to finish the work.

Mr. Loudon had previously requested Dr. JAMIESON
to write the chapters on Mechanics, Hydrostatics, and
Hydraulics, which that gentleman has done; and Mr. JAY,
a friend of his, has kindly furnished the section on Farm
Book-keeping.   These are the only portions that were
never seen by Mr. Loudon.   Some weeks before his
death, the chapters on Architectural Drawing and Isometrical Projection and Perspective were written for him
by Mr. ROBERTSON, and contributions for the work were
also sent to him by SAMUEL TAYLOR, Esq., RICHARD
VARDEN, Esq., Professor DONALDSON, Mr. JAMES
MUNRO, and some other kind friends.   To all these gentlemen I here offer my sincere thanks and acknowledge-
ments, satisfied that in so doing I am only acting as my
husband would have done, had his life been spared to the
present time.

J. W. LOUDON.

Bayswater, September 30. 1845.

CONTENTS.

A 3

A 4

CHAPTER X.

CHAPTER XI.

CHAPTER XII.

CHAPTER XIII.

MISCELLANEOUS TABLES.

VIGNETTES.

ACCOUNT OF THE LIFE AND WRITINGS

JOHN CLAUDIUS LOUDON.

JOHN CLAUDIUS LOUDON was born on the 8th of April, 1783, at Cambuslang, in Lanarkshire, the residence of his mother's only sister, herself the mother of Dr. Claudius Buchanan (the author of a work entitled *Christian Researches in Asia*), whose labours in India, in attempting to convert and instruct the Hindoos, have made his name celebrated in the religious world. Mr. Loudon was the eldest of a large family; and his father, who was a farmer, residing at Kerse Hall, near Gogar, about five miles from Edinburgh, being a man of enlightened mind and superior information, was very anxious that he should have every possible advantage in his education. Strange to say, however, Mr. Loudon, when a boy, though fond of books, had an insuperable aversion from learning languages, and no persuasions could induce him to study Latin and French, though his father had a master from Edinburgh purposely to teach him the latter language. At this early period, however, a taste for landscape-gardening began to show itself, as his principal pleasure was in making walks and beds in a little garden his father had given him; and so eager was he to obtain seeds to sow in it, that, when a jar of tamarinds arrived from an uncle in the West Indies, he gave the other children his share of the fruit, on condition

of his having all the *seeds*.   While yet quite a child, he
was sent to live with an uncle in Edinburgh, that he might
attend the classes at the public schools.   Here he over-
came his dislike to Latin, and made extraordinary progress
in drawing and arithmetic.   He also attended classes of
botany and chemistry, making copious notes, illustrated
with very clever pen-and-ink sketches.   Still he could not
make up his mind to learn French, till one day, when he
was about fourteen, his uncle, showing a fine French
engraving to a friend, asked his nephew to translate the
title.   This he could not do; and the deep shame and mor-
tification which he felt, and which he never afterwards
forgot, made him determine to acquire the language.
Pride, however, and a love of independence, which was
ever one of his strongest feelings, prevented him from
applying to his father to defray the expense; and he actually
paid his master himself, by the sale of a translation which
he afterwards made for the editor of a periodical then publishing in Edinburgh.   He subsequently studied Italian,
and paid his master in the same manner.   He also kept a
Journal from the time he was thirteen, and continued it for
nearly thirty years; writing it for many years in French,
in order to familiarise himself with the language.

Among all the studies which Mr. Loudon pursued
while in Edinburgh, those he preferred were writing and
drawing.   The first he learned from Mr. Paton, afterwards
father to the celebrated singer of that name; and strange
enough, I have found an old letter of his to Mr. Loudon, sen., prophesying that his son John would be one of
the best writers of his day — a prophecy that has been
abundantly realised, though certainly not in the sense
its author intended it.   Drawing was, however, his
favourite pursuit; and in this he made such proficiency,
that, when his father at last consented to his being brought
up as a landscape-gardener, he was competent to take
the situation of draughtsman and assistant to Mr. John
Mawer, at Easter Dalry, near Edinburgh.   Mr. Mawer
was a nurseryman, as well as a planner (as the Scotch

call a landscape-gardener); and, while with him, Mr. Loudon learned a good deal of gardening generally, particularly of the management of hothouses. Unfortunately, Mr. Mawer died before his pupil was sixteen; and for three or four years afterwards Mr. Loudon resided with Mr. Dickson, a nurseryman and planner in Leith Walk, where he acquired an excellent knowledge of plants. There he boarded in Mr. Dickson's house; and, though remarkable for the nicety of his dress, and the general refinement of his habits, his desire of improvement was so great, that he regularly sat up two nights in every week to study, drinking strong green tea to keep himself awake; and this practice of sitting up two nights in every week he continued for many years. While at Mr. Dickson's, he attended classes of botany, chemistry, and agriculture; the last under Dr. Coventry, who was then Professor of Agriculture in the University of Edinburgh, and he was considered by that gentleman to be his most promising pupil.

In truth, it has been highly gratifying to me, while turning over family papers to obtain what particulars I could of my husband's early life, to find continually, in old copy and account books, letters which had been no doubt treasured up by his mother, from different persons under whom he had studied, bearing the most honourable testimony to his proficiency in the various branches of his education, and particularly noting his unwearied perseverance in making himself thoroughly master of whatever he undertook. Mr. Loudon was not a man of many words, and he was never fond of showing the knowledge he possessed; but it was astonishing how much he did know on every subject to which he had turned his attention.

In 1803 he first arrived in London. The following day he called on Mr. Sowerby, Mead Place, Lambeth, who was the first gentleman he visited in England; and he was exceedingly delighted with the models and mineralogical specimens, which were so admirably arranged as to give

him the greatest satisfaction from his innate love of order;
and he afterwards devised a plan for his own books and
papers, partly founded on that of Mr. Sowerby, but much
more complete.

As he brought a great number of letters of recommendation to different noblemen and gentlemen of landed
property, many of them being from Dr. Coventry with
whom he was a great favourite, he was soon extensively employed as a landscape-gardener; and his journal
is filled with accounts of his tours in various parts of
England. It is curious, in turning over his memoranda,
to find how many improvements suggested themselves to
his active mind, which he was unable, from various circumstances, to carry into effect at the time, but which, many
years afterwards, were executed either by himself or by
other persons, who, however, were unaware that he had
previously suggested them. Throughout his life similar
occurrences were continually taking place; and nothing
was more common than for him to find persons taking the
merit to themselves of inventions which he had suggested years before. When this happened, he was frequently urged to assert his prior claim; but he always
answered, that he thought the person who made an invention useful to the public had more merit than its original
contriver; and that, in fact, so long as the public were
benefited by any invention of his, it was perfectly indifferent to him who had the merit of it. There never lived
a more liberal and thoroughly public-spirited man than
Mr. Loudon. He had not a single particle of selfishness
in his disposition, and in all his actions he never took
the benefit they would produce to himself into consideration. When writing a book, his object was to obtain the
best possible information on the subject he had in hand;
and he was never deterred from seeking this by any considerations of trouble or expense.

That these feelings influenced him from the time of his
first arrival in England may be traced in every page
of his Journal; and that they continued to influence him

to the last day of his life was only too evident to every one around him at that mournful period.

When Mr. Loudon first arrived in London, he was very much struck with the gloomy appearance of the gardens in the centre of the public squares, which were then planted almost entirely with evergreens, particularly with Scotch pines, yews, and spruce firs; and, before the close of the year 1803, he published an article in a work called *The Literary Journal,* which he entitled, " Observations on laying out the Public Squares of London." In this article he blamed freely the taste which then prevailed, and suggested the great improvement that would result from banishing the yews and firs (which always looked gloomy from the effect of the smoke on their leaves), and mingling deciduous trees with the other evergreens. He particularly named the Oriental and Occidental plane trees, the sycamore, and the almond, as ornamental trees that would bear the smoke of the city; and it is curious to observe how exactly his suggestions have been adopted, as these trees are now to be found in almost every square in London.

About this time he appears to have become a member of the Linnean Society, probably through the interest of Sir Joseph Banks, to whom he had brought a letter of introduction, and who, till his death in 1820, continued his warm friend. At the house of Sir Joseph Banks Mr. Loudon met most of the eminent scientific men of that day, and the effect produced by their conversation on his active mind may be traced in his Journal. Among many other interesting memoranda of new ideas that struck him about this period, is one as to the expediency of trying the effects of charcoal on vegetation, from having observed the beautiful verdure of the grass on a spot where charcoal had been burnt. He appears, however, to have thought no more at that time on the subject, or to have forgotten it, as, when he afterwards wrote on charcoal, he made no allusion to this fact.

In 1804, having been employed by the Earl of Mansfield to make some plans for altering the Palace Gardens at Scone in Perthshire, he returned to Scotland and remained there several months, laying out grounds for many noblemen and gentlemen. While thus engaged, and while giving directions for planting and managing woods, and on the best mode of draining and otherwise improving estates, several ideas struck him, which he afterwards embodied in a book published in Edinburgh by Constable and Co., and by Longman, Hurst, Rees, and Orme, in London. This, then, was the first work of Mr. Loudon's presented to the public through the Messrs. Longman, with whom he continued to transact business of the same nature for nearly forty years. The book alluded to was entitled *Observations on the Formation and Management of Useful and Ornamental Plantations; on the Theory and Practice of Landscape-Gardening, and on gaining and embanking Land from Rivers or the Sea.* As this was his first separate work, and as it is now comparatively little known, it may be interesting to copy a few sentences from the Introduction; which will show how strongly his mind was, even in his youth, imbued with the subject of his profession, though he was then apparently disposed to treat it in a different style from what he did in after years.

" Various are the vegetable productions which this earth affords. Blades of grass spring up every where, and clothe the surface with pasture; groups of shrubs arise in some places, and diversify this uniform covering; but trees are the most striking objects that adorn the face of inanimate nature. If we imagine for a moment that the surface of Europe were totally divested of wood, what would be our sensations on viewing its appearance? Without this accompaniment, hills and valleys, rivers and lakes, rocks and cataracts, all of themselves the most perfect that could be imagined, would present an aspect bleak, savage, and uninteresting. But, let the mountains be covered with wood, and the water shaded by trees, and the scene is instantly changed: what was before cold and barren, is now rich, noble, and full of variety. In travelling through a naked country, a whole unvaried horizon is comprehended by the eye with a single glance; its surface is totally destitute of intricacy to excite curiosity and fix attention; and both the eye and the mind are kept in a state of perpetual weariness and fatigue. But, in a wooded country, the scene is continually changing; the trees form a varied boundary to every thing around, and enter into numberless and pleasing combinations with all other objects; the eye is relieved without distraction, and the mind fully engaged without fatigue. If we examine even a tree by itself, the intricate formation and disposition of its boughs, spray, and leaves, its varied form, beautiful tints, and diversity of light and shade, make it far surpass every other object; and, notwithstanding this multiplicity of separate parts, its general effect is simple and grand.

" But wood is not only the greatest ornament on the face of our globe, but the most essential requisite for the accommodation of civilised society. The implements of agriculture, the machinery of manufactures, and the vehicles of commercial intercourse, are all made of timber; nor is there an edifice or superstructure of almost any denomination, in which this material does not form the principal part.

" Wood is more particularly valuable in Great Britain, where the existence and prosperity of the empire depends upon the support of a numerous shipping, emphatically called its ' wooden walls.'

" From the universal utility, and the unrivalled beauty of wood, it may reasonably be supposed to have been assiduously cultivated in all improved countries; and, accordingly, we find trees were planted, and the growth of timber encouraged, by every polished nation. To this subject, as to all other parts of rural economy, the Romans paid great attention; and the writings of some of their most celebrated authors contain many excellent observations and precepts on the culture and management of timber and ornamental trees." (p. 20.)

" But, independently of the beauty and profit of wood, the pleasure attending the formation and management of plantations will be a considerable recommendation to every virtuous mind. We look upon our young trees as our offspring; and nothing can possibly be more satisfying than to see them grow and prosper under our care and attention; nothing more interesting than to examine their progress, and mark their several peculiarities. As they advance to perfection, we foresee their ultimate beauty; and the consideration that we have reared them raises a most agreeable train of sensations in our minds; so innocent and rational, that they may justly rank with the most exquisite of human gratifications. But, as the most powerful motives to planting are those which address themselves to the interest of the individual, I proceed to consider it more particularly in this point of view." (p. 23.)

The work is divided into sections, in one of which, in particular, on the principal distinctions of trees and shrubs, are some very interesting observations, which show how well their author was acquainted with the characteristics of trees and shrubs even at that early period of his life. Before Mr. Loudon left Edinburgh, he published another work, entitled *A short Treatise on some Improvements lately made in Hothouses.* This was in 1805; and the same year he returned to England. On this second voyage to London, he was compelled by stress of weather to land at Lowestoffe; and he took such a disgust at the sea, that he never afterwards travelled by it if it was possible to go by land. He now resumed his labours as a landscape-gardener; and his Journal is filled with the observations he made, and the ideas that suggested themselves of improvements, on all he saw. Among other things, he made some remarks on the best mode of harmonising colours in flower-gardens, which accord, in a very striking manner, with the principles afterwards laid down by M. Chevreul in his celebrated work entitled *De la Loi du Contraste simultané des Couleurs,* published in Paris in 1839. Mr. Loudon states that he had observed that flower-gardens

looked best when the flowers were so arranged as to
have a compound colour next the simple one, which was
not contained in it.   Thus, as there are only three simple
colours, blue, red, and yellow, he advises that purple
flowers, which are composed of blue and red, should have
yellow next them; that orange flowers, which are composed of red and yellow, should be contrasted with blue;
and that green flowers, which are composed of blue and
yellow, should be relieved by red.   He accounts for this
on the principle that three parts are required to make a
perfect whole; and he compares the union of the three
primitive colours formed in this manner with the common
chord in music; an idea which has since been worked out
by several able writers.   He had also formed the plan of
a Pictorial Dictionary, which was to embrace every kind
of subject, and to be illustrated by finished woodcuts
printed with the type.

In 1806 Mr. Loudon published his *Treatise on forming,
improving, and managing Country Residences, and on the
Choice of Situations appropriate to every Class of Pur-
chasers.   With an Appendix containing an Enquiry into
the Utility and Merits of Mr. Repton's Mode of showing
Effects by Slides and Sketches, and Strictures on his Opi-
nions and Practice in Landscape-Gardening.   Illustrated by
Descriptions of Scenery and Buildings, by References to
Country Seats and Passages of Country in most Parts of
Great Britain, and by 32 Engravings.*

This work was much more voluminous than any of the
preceding ones, and it was ornamented by some elegant
copperplate engravings of landscape scenery, drawn by
himself, which, in 1807, were republished, with short descriptions, as a separate work.

During the greater part of the year 1806 Mr. Loudon
was actively engaged in landscape-gardening; and towards
the close of that year, when returning from Trè-Madoc,
in Caernarvonshire, the seat of W. A. Madocks, Esq.,
he caught a violent cold by travelling on the outside of a
coach all night in the rain, and neglecting to change his

clothes when he reached the end of his journey. The cold brought on a rheumatic fever, which settled finally in his left knee, and, from improper medical treatment, terminated in a stiff joint; a circumstance which was a source of great annoyance to him, not only at the time when it occurred, but during the whole of the remainder of his life. This will not appear surprising, when it is considered that he was at that period in the prime of his days, and not only remarkably healthy and vigorous in constitution, but equally active and independent in mind. While suffering from the effects of the complaint in his knee, he took lodgings at a farm-house at Pinner, near Harrow; and, while there, the activity of his mind made him anxiously enquire into the state of English farming. He also amused himself by painting several landscapes, some of which were exhibited at the Royal Academy, and by learning German, paying his expenses, as he had done before when he learned French, by selling for publication a pamphlet which he had translated by way of exercise. In this case, the translation being of a popular work, it was sold to Mr. Cadell for 15*l*. He also took lessons in Greek and Hebrew. The following extract from his Journal in 1806 will give some idea of his feelings at this period : — " Alas! how have I neglected the important task of improving myself! How much I have seen, what new ideas have developed themselves, and what different views of life I have acquired since I came to London three years ago! I am now twenty-three years of age, and perhaps one third of my life has passed away, and yet what have I done to benefit my fellow men ?"

Mr. Loudon, during the length of time he was compelled to remain at Pinner, became so interested respecting English farming, and so anxious that the faults he observed in it should be corrected, that he wrote to his father, stating the capability of the soil, and the imperfect state of the husbandry, and urging him to come to England. It happened that at this period the farm called Wood Hall, where he had been staying so long, was to be let, and Mr. Loudon, senior, in consequence of the recommendation of his son, took it, and removed to it in 1807. The following year Mr. Loudon, who was then residing with his father at Wood Hall, wrote a pamphlet entitled *An immediate and effectual Mode of raising the Rental of the Landed Property of England; and rendering Great Britain independent of other Nations for a Supply of Bread Corn. By a Scotch Farmer, now farming in Middlesex.* This pamphlet excited a great deal of attention; and General Stratton, a gentleman possessing a large landed estate, called Tew Park, in Oxfordshire, having read it, was so much interested in the matter it contained, that he offered him a portion of his property at a low rate, in order that he might undertake the management of the rest, and thus introduce Scotch farming into Oxfordshire.

The farm which Mr. Loudon took from General Stratton, and which was called Great Tew, was nearly eighteen miles from the city of Oxford, and it contained upwards of 1500 acres. " The surface," as he describes it, " was diversified by bold undulations, hills, and steeps, and the soil contained considerable variety of loam, clay, and light earth, on limestone and red rock. It was, however, subdivided in a manner the most unsuitable for arable husbandry, and totally destitute of carriage roads. In every other respect it was equally unfit for northern agriculture, having very indifferent buildings, and being greatly in want of draining and levelling." At this place he established a kind of agricultural college for the instruction of young men in rural pursuits; some of these, being the sons of landed proprietors, were under his own immediate superintendence; and others, who were placed in a second class, were instructed by his bailiff, and intended for land-stewards and farm-bailiffs. A description of this college, and of the improvements effected at Great Tew, was given to the public in 1809, in a pamphlet entitled *The Utility of Agricultural Knowledge to the Sons of the Landed Proprietors of England, and to Young Men intended for Estate-Agents; illustrated by what has taken place in Scotland. With an Account of an Institution formed for Agricultural Pupils in Oxfordshire. By a Scotch Farmer and Land-Agent, resident in that County.* In this pamphlet there is one passage showing how much attached he was to landscape-gardening, an attachment which remained undiminished to his death; and how severely he felt the misfortune of having his knee become anchylosed from the effects of the rheumatic fever before alluded to. The passage, which occurs in the introductory part of the work, is as follows: — " A recent personal misfortune, by which the author incurred deformity and lameness, has occasioned his having recourse to farming as a permanent source of income, lest by any future attack of disease he should be prevented from the more active duties and extensive range of a beloved profession on which he had formerly been chiefly dependent."

Notwithstanding the desponding feelings expressed in this paragraph, Mr. Loudon appears from his memorandum books to have been still extensively engaged in landscape-gardening, as there are memoranda of various places that he laid out in England, Wales, and Ireland, till the close of 1812. Before this period he had quitted Tew; and finding that he had amassed upwards of 15,000*l.* by his labours, he determined to relax his exertions, and to gratify his ardent thirst for knowledge by travelling abroad. Previously, however, to doing this, he published two works: one entitled *Hints on the Formation of Gardens and Pleasure-Grounds, with Designs in various Styles of Rural Embellishment: comprising Plans for laying out Flower, Fruit, and Kitchen Gardens; and the Construction and Arrangement of Glass Houses, Hot Walls, and Stoves; with Directions for the Management of Plantations, and a Priced Catalogue of Fruit and Forest Trees, Shrubs, and Herbaceous Plants; the whole adapted to Villa Grounds from one Perch to One Hundred Acres in Extent:* and the other, *Observations on laying out Farms in the Scotch Style adapted to England.*

The first of these works I have no copy of, and have

never seen; but the second is now before me, and it contains many interesting particulars respecting the farm of Great Tew rented by himself, and those of Wood Hall and Kenton Lane rented by his father. From this work it appears, that, though Mr. Loudon, senior, enjoyed but a few months' health after settling at Wood Hall, which he entered upon at Michaelmas, 1807, his death taking place in December, 1809, the estate was so much improved, even in that short period, that it was let after his death for a thousand pounds a year, being three hundred pounds a year more than he had paid for it. It also appears that Mr. Loudon entered on the farm at Great Tew at Michaelmas, 1808, and left it in February, 1811; General Stratton paying him a considerable sum for his lease, stock, and the improvements he had effected.

The Continent, after having been long closed to English visitors, was thrown open in 1813 by the general rising against Napoleon Bonaparte, and it presented an ample field to an enquiring mind like that of Mr. Loudon. Accordingly, after having made the necessary preparations, he sailed from Harwich on the 16th of March. He first landed at Gottenburg, and was delighted with Sweden, its roads, its people, and its systems of education; but he was too impatient to visit the theatre of war to stay long in Sweden, and he proceeded by way of Memel to Konigsberg, where he arrived on the 14th of April. In this country he found every where traces of war: skeletons of horses lay bleaching in the fields, the roads were broken up, and the country houses in ruins. At Elbing he found the streets filled with the goods and cattle of the country people, who had poured into the town for protection from the French army, which was then passing within two miles of it; and near Marienburg he passed through a bivouac of 2000 Russian troops, who in their dress and general appearance looked more like convicts than soldiers. The whole of the valley between Marienburg and Dantzic he found covered with water, and looking like one vast lake; but on the hills near Dantzic there was an encampment of Russians; the

Cossacks belonging to which were digging holes for themselves and horses in the loose sand. These holes they
afterwards covered with boughs of trees, stuck into the
earth and meeting in the centre as in a gipsy tent; the
whole looking, at a little distance, like a number of huts of
the Esquimaux Indians. He now passed through Swedish
Pomerania; and, on approaching Berlin, found the long
avenues of trees leading to that city filled with foot passen-
gers, carriages full of ladies, and waggons full of luggage,
all proceeding there for protection; and forming a very
striking picture as he passed through them by moonlight.

He remained at Berlin from the 14th of May to the 1st
of June, and then proceeded to Frankfort on the Oder.
Here, at the table d'hôte, he dined with several Prussian
officers, who, supposing him to be a Frenchman, sat for
some time in perfect silence: but, on hearing him speak
German, one said to the other, " He must be English;"
and, when he told them that he came from London, they
all rose, one springing over the table in his haste, and
crowded round him, shaking hands, kissing him, and over-
whelming him with compliments, as he was the first Eng-
lishman they had ever seen. He then proceeded through
Posen to Warsaw, where he arrived on the 6th of June.

Afterwards he travelled towards Russia, but was stopped
at the little town of Tykocyn, and detained there three
months, from some informality in his passport. When this
difficulty was overcome, he proceeded by Grodno to Wilna,
through a country covered with the remains of the French
army, horses and men lying dead by the road-side, and
bands of wild-looking Cossacks scouring the country. On
entering Kosnow three Cossacks attacked his carriage,
and endeavoured to carry off the horses, but they were
beaten back by the whips of the driver and servants. At
Mitton he was obliged to sleep in his britzska, as every
house was full of the wounded; and he was awakened in
the night by the cows and other animals, of which the inn
yard was full, eating the hay which had been put over his
feet to keep them warm. He reached Riga on the 30th

of September, and found the town completely surrounded
by a barricade of waggons, which had been taken from
the French. Between this town and St. Petersburg, while
making a drawing of a picturesque old fort, he was taken
up as a spy; and, on his examination before the prefect,
he was much amused at hearing the comments made on
his note-book, which was full of unconnected memoranda,
and which puzzled the magistrates and their officers ex-
cessively when they heard it translated into Russ.

Mr. Loudon reached St. Petersburg on the 30th of October, just before the breaking up of the bridge, and he remained there three or four months; after which he proceeded
to Moscow, where he arrived on the 4th of March, 1814,
after having encountered various difficulties on the road.
Once, in particular, the horses in his carriage being unable
to drag it through a snow-drift, the postilions very coolly
unharnessed them and trotted off, telling him that they
would bring fresh horses in the morning, and that he
would be in no danger from the wolves, if he would keep
the windows of his carriage close, and the leather curtains
down. There was no remedy but to submit; and few
men were better fitted by nature for bearing the horrors of
such a night than Mr. Loudon, from his natural calmness
and patient endurance of difficulties. He often, however,
spoke of the situation he was in, particularly when he
heard the howling of the wolves, and once when a herd of
them rushed across the road close to his carriage. He had
also some doubts whether the postilions would be able to
recollect where they had left the carriage, as the wind had
been very high during the night, and had blown the snow
through the crevices in the curtains. The morning, however, brought the postilions with fresh horses, and the remainder of the journey was passed without any difficulty.

When he reached Moscow, he found the houses yet
black from the recent fire, and the streets filled with
the ruins of churches and noble mansions. Soon after his
arrival news was received of the capture of Paris, and the
entrance of the allied sovereigns into that city; but the

Russians took this intelligence so coolly, that, though it reached Moscow on the 25th of April, the illuminations in honour of it did not take place till the 5th of May. He left Moscow on the 2d of June, and reached Kiov on the 15th. Here he had an interview with General Rapp on account of some informality in his passport. He then proceeded to Cracow, and thence to Vienna; after which he visited Prague, Dresden, and Leipsic, passing through Magdeburg to Hamburg, where he embarked for England, and reached Yarmouth on the 27th of September, 1814.

During this long and interesting journey Mr. Loudon visited and took views of nearly all the palaces and large rural residences in the countries through which he passed; and he visited all the principal gardens, frequently going two or three days' journey out of his route, if he heard of any garden that he thought worth seeing. He also visited most of the eminent scientific men in the different cities he passed through; and was elected a member of the Imperial Society of Moscow, the Natural History Society at Berlin, the Royal Economical Society at Potsdam, and many others. I have often wondered that on his return home he did not publish his travels; as the Continent was then, comparatively, so little known, that a a narrative of what he saw, illustrated by his sketches, would have been highly interesting. Business of a very unpleasant nature, however, awaited him, and probably so completely occupied his mind as to leave no room for any thing else.

I have already mentioned that when Mr. Loudon went abroad he had a large sum of money lying unemployed in his banker's hands; and with this he was induced, I know not how, to embark in mercantile speculations, and underwriting ships at Lloyd's. As he knew nothing of business of this nature, it is not surprising that his speculations turned out badly; and for more than twelve months he was involved in pecuniary difficulties. I am unable to give all the details of his sufferings during this period, as it was a subject he never spoke of, and the allusions to it

in his memorandum books are by no means explicit.   It
appears, however, that, after having made several fruitless
journeys (including one to Paris in 1815) in the hope of
recovering some part of the property, he was compelled to
submit to the loss of nearly the whole; and that his health
was very seriously injured by the anxieties he underwent.

About this time (1816) his mother and sisters left
the country; and he, having determined that in future
they should reside with him, took a house at Bayswater
called the Hermitage, which had a large garden annexed.
His health was now seriously impaired, but his mind
always seemed to acquire additional vigour from the
feebleness of his body; and, as he was unable to use so
much exertion as he had formerly done in landscape-gardening, he amused himself by trying experiments relating
to the construction of hothouses, and by having several of
different kinds erected in his garden.

In August, 1815, a paper had been published in the
Transactions of the Horticultural Society, by Sir George
Mackenzie of Coul, " on the form which the glass of a
forcing-house ought to have, in order to receive the greatest
possible quantity of rays from the sun."   This form Sir
George conceived to be that of a globe; but, as it seemed
impracticable to make a hothouse globular, he proposed
to make the roof the segment of a circle.   Mr. Loudon
appears to have been very much struck with this paper;
but he saw faults in the plan which he thought might be
amended, and he tried houses with curvilinear roofs of
various kinds, in order to ascertain which was the best.
He also tried a house with what he called ridge and furrow glazing; a plan which has since been carried out on a
magnificent scale by Mr. Paxton, in the Duke of Devonshire's splendid conservatory at Chatsworth.   While these
houses were in progress, he wrote a work entitled *Remarks on the Construction of Hothouses*, &c., which was
published in 1817.   Shortly afterwards he invented a
new kind of sash-bar, of which he gave a description,
together with sketches of the hothouses, and details of

their construction, in a quarto pamphlet entitled *Sketches
of Curvilinear Hothouses*, &c., which was published in 1818.
The profits of this bar he was to have shared with the iron-
monger by whom it was sold; but, I believe, he never
reaped any pecuniary advantage from it. He also published, in folio, another work, in the same year, entitled *A
Comparative View of the Common and Curvilinear Modes
of roofing Hothouses.*

He now seems to have determined on devoting his time
principally to his pen; and he began to collect materials
for the well-known *Encyclopædia of Gardening.* It is
probable that the first idea of this work had occurred
to him while he was travelling, from the great number of
gardens he had seen, and the various modes of gardening
that he had found practised in different countries. At any
rate, he determined to commence his work with a history
of gardening, and a description of the gardens of various
countries; introducing illustrative drawings engraved on
wood and printed with the text, this being, I believe,
the first time any engravings, except mere outlines, had
been printed in that manner. It was necessary, in
order to complete his plan, that he should see the gardens
of France and Italy, in the same manner as he had seen
those of the North of Europe; and, for this purpose, he
determined to set out on another tour, though his health
was at that time so very indifferent, that one of his
friends, who saw him at Dover, told him he looked more
fit to keep his bed than to set out on a journey. Mr.
Loudon, however, was not easily deterred from any thing
that he had resolved upon, and he proceeded by way of
Calais and Abbeville to Paris, where he arrived on the
30th of May, 1819. After seeing every thing deserving
of notice in Paris, and becoming acquainted with many
eminent men there, from the letters of introduction given
to him by his kind friend Sir Joseph Banks, he left on the
10th of June for Lyons; in the Botanic Garden of which
city he saw for the first time a living plant of the Vallisnèria, which had not then been introduced into England,

and which he had only seen in a dry state in the Hortus
Siccus of Sir Joseph Banks. From Lyons he went to
Avignon, and then he visited the celebrated fountain of
Vaucluse.   Afterwards he proceeded to Marseilles, and
thence to Nice, from which city he sailed in a felucca for
Genoa.

During the whole of his tour through France he visited
the gardens every where, and made memoranda of every
thing that he thought would be useful for his intended
work.   He also made sketches of all the principal places,
as he had previously done in the North of Europe.

Before leaving Genoa he procured a collection of orange
trees, which he sent to England for his greenhouse at
Bayswater.   He also saw, for the first time, slate boxes
used for orange trees, in the garden of Signore di Negre,
near Genoa.   In this city, also, he first met with his
friend Captain Mangles; and, joining him and Captain
Irby, they travelled together along the shores of the
Mediterranean, leaving Genoa on the 6th of July in a
felucca for Leghorn, where they arrived on the 8th, and
thence proceeded through Pisa to Florence.   During the
whole of this tour Mr. Loudon's Journal is entirely filled
with descriptions of the gardens he visited, observations on
the different modes of culture he saw practised, and various
remarks on the habits of plants.   One of the latter, which
appears to me worth recording, is, that he found *Saxífraga
crassifòlia* killed by a very slight frost in Florence; though
it will bear a considerable degree of cold in more northern
climates.   From Florence he went to Rome, and thence
to Naples; after which he visited Pompeii and Hercula-
neum, returning through Rome to Florence on the 21st of
August.   In these cities he visited all that is generally
considered worth seeing; and, of course, did not neglect
his favourite gardens.

About this period he saw for the first time a specimen
of the trick often practised by the Italian gardeners,
which is called by the French *Greffe des Charlatans*.   This
consists in taking the pith out of the trunk and branches

of an orange tree, and dexterously introducing through these a rose tree, or any other plant which it is wished shall appear to have been grafted on the orange. Care is taken not to injure the roots of either; and, if put cautiously into the ground, both will produce leaves and flowers.

The next place he visited was Bologna, near which he passed a day or two with an Italian family who were enjoying the pleasures of the vintage. He then went through Ferrara to Venice; the first part of the road to which was bordered by hedges, in which were vines laden with grapes hanging from tree to tree. At Deux Ponts, he embarked in a boat, and found the canal nearly all the way to Venice full of beautiful aquatic plants, among which was the Vallisnèria. He was very much struck with the imposing view that he first obtained of Venice, including the grand square of St. Mark, with its winged lion on a granite column. He also remarked the freshness and brilliancy of the paintings; and he noticed that the Post-office at Venice was built upon immense piles of logwood. The whole of the first night that he passed in Venice he was unable to sleep, from the number of persons that were singing in parties in the streets. The following morning he hired a gondola, and went through the city, with which he was exceedingly delighted; for, as he says, emphatically, " It is impossible to know what Italian architecture and Italian paintings really are, without seeing those at Venice." Before leaving this splendid city, he procured a living plant of the Vallisnèria, which he placed in a little tin can containing water, and carried himself, when he was travelling, lest any harm should happen to it.

The next place he visited was Padua, where he saw the celebrated Botanic Garden. The road from this to Vicenza was bordered with hedges of *H*ibíscus syrìacus. He had now entered upon the district where silk is chiefly produced, and found on each side of the road vast plantations of white mulberry trees. Thence he proceeded to Milan ; after which he visited the splendid gardens of Monza, with which he was most exceedingly delighted. He found here square pots universally used for the plants in the greenhouses, in order to save room; and the tubs of the orange and lemon trees sunk in the ground, to keep the plants moist. He found the tuberoses most luxuriant, and scenting the air. The Botanic Garden at Milan is small but well filled. On leaving Milan he visited the Borromean Isles; but thought the beauty of Isola Bella somewhat exaggerated.

The little can containing the Vallisnèria had occasioned him a great deal of trouble during his journey through the North of Italy; and he found it still more difficult to take care of while he was crossing the Simplon into Switzerland, as he was obliged to perform the journey on a mule. However, to use his own expression, he nursed it as carefully as he would have done a child, and the Vallisnèria was in perfect health when he arrived at Geneva on the 13th of September, 1819. Here he visited the Botanic Garden, and formed an acquaintance with the late Professor De Candolle. He afterwards visited Basle; saw the establishment of M. Fellenberg, and proceeded through Strasburg to Paris, where he only slept one night, and then set off for Belgium. The one night that he passed at Paris proved unfortunately fatal to the Vallis-nèria. The inn he went to happened to be crowded when he arrived, and he was placed in a very small bedroom, that was so hot and close he fancied his poor plant looked drooping. To revive it, he opened the window, and placed the tin can on the window-sill, taking great care to secure it that it might not fall. In the morning, however, though the tin can remained, the plant was gone; and he was never able to ascertain what had become of it, though he supposed it had been carried off by sparrows.

At Brussels he found the Botanic Garden in those days nothing; but he liked the park and the promenade on the ramparts, to which the Botanic Garden has since been removed At Ghent, he was also much pleased with the Botanic Garden, and with the generally luxuriant appear-ance of the plants in the private gardens near the town.

In Bruges and Ostend he found little to see; and he returned to Bayswater on the 9th of October.

As soon as he reached home, he began the *Encyclopædia of Gardening,* at which he worked with little intermission till it was finished, though he was suffering severely at the time from chronic rheumatism in his right arm; the pain from which became at length so intolerable, that in 1820 he was compelled to call in medical aid; and, being recommended to try Mahomed's vapour baths, he went down to Brighton for that purpose. Here, notwithstanding the extreme torture he suffered from the shampooing and stretching, he submitted to both with so much patience, that they were continued by the operators till they actually broke his right arm so close to the shoulder as to render it impossible to have it set in the usual manner, and consequently it never united properly, though he continued to use his hand to write with for several years.

In 1822 appeared the first edition of the *Encyclopædia of Gardening;* a most laborious work, remarkable both for the immense mass of useful matter it contains, and for the then unusual circumstance of a great number of finished wood-engravings being printed with the text, instead of being in separate pages. This book had an extraordinary sale, and fully established the literary fame of its author.

In the early part of the year 1823 he wrote a work entitled *The different Modes of cultivating the Pine-apple, from its first Introduction to Europe to the Improvements of T. A. Knight, Esq., in* 1822.

About this time also a little work was published anonymously, called *The Greenhouse Companion,* which, I believe, was written, either entirely or in part, by Mr. Loudon: but it must have been by a wonderful exertion, if he did write it; as during the whole of the year 1823 he suffered most excruciating pain, not only from his right arm, the bone of which had never properly united, and to retain which in its place he was compelled to wear an iron case night and day, but from the rheumatism which had settled in his left hand, and which contracted two of his fingers and his thumb, so as to render them useless. It is, however, worthy of remark, and quite characteristic of Mr. Loudon, that, at the very time he was suffering such acute bodily pain, he formed the plan of his houses in Porchester Terrace, Bayswater, and superintended the building of them himself, rising at four o'clock every morning, that he might be on the spot when the workmen came to their work.

In 1824 a second edition was published of the *Encyclopædia of Gardening;* in which the work was nearly all rewritten, and very considerable additions were made to it. In the following year, 1825, the *Encyclopædia of Agriculture* was written and published. These extensive and laborious works following closely upon each other, in Mr. Loudon's state of health, speak strongly as to his unparalleled energy of mind. When, shortly after, his right arm was broken a second time, and he was obliged to submit to amputation, though he gave up landscape-gardening, it was only to devote himself more assiduously to his pen. He was, however, now no longer able to write or draw himself, and he was compelled to employ both an amanuensis and a draughtsman. Still, though he had only the use of the third and little finger of his left hand, he would frequently take a pen or a pencil, and make sketches with astonishing vigour, so as fully to explain to his draughtsman what he wished to be done.

During the time that he was suffering so severely from the pain in his arm, he found no ease but from taking laudanum; and he became at last so habituated to the use of this noxious potion, that he took a wine-glassful every eight hours. After the amputation of his arm, however, he wished to leave off taking it, as he was aware of its injurious effects upon his general health; and he contrived to cure himself by putting a wine-glassful of water into his quart bottle of laudanum every time he took out a wine-glassful of the potion, so that the mixture became gradually weaker every day, till at last it was little more than water; and he found he had cured himself of this dangerous habit without experiencing any inconvenience.

In 1826 he established *The Gardener's Magazine,* the first periodical devoted exclusively to horticultural subjects. This work was always Mr. Loudon's favourite, and the organ through which he communicated his own thoughts and feelings to the public. It was originally undertaken principally for the benefit of gardeners in the country, in order to put them " on a footing with those about the metropolis;" but it soon became the universal means of communication among gardeners, and was of incalculable benefit to them. It also became a source of great pleasure to amateurs of gardening, and was no doubt the means of inspiring a taste for the pursuit in many who had before been indifferent to it. " In an art so universally practised as gardening, and one daily undergoing so much improvement," Mr. Loudon observes, " a great many occurrences must take place worthy of being recorded, not only for the entertainment of gardening readers, but for the instruction of practitioners in the art." (*Gard. Mag.* vol. i. p. 1.) That this work met the wants of a large class of readers is evident from four thousand copies of the first number having been sold in a few days; and from the work having continued popular for nineteen years, and, in fact, till its close at the death of its conductor.

The Gardener's Magazine first appeared quarterly, afterwards it was published every two months, and finally every month. The second number of this work contained an attack on the London Horticultural Society, the affairs of which were then notoriously ill managed, though before the publication of *The Gardener's Magazine* no one had ventured to complain of them publicly. In the same number appeared an article on the " Self-education of Gardeners;" in which Mr. Loudon began those earnest exhortations to gardeners to improve themselves, and those efforts to put them in the way of self-improvement, which he continued almost to the last hour of his life. He also, in this second number, gave a plan for the improvement of

Kensington Gardens, and suggested the erection of "small stone lodges with fireplaces at the principal garden gates, for the comfort of the door-keepers in winter," as before that time the door-keepers had no shelter but the alcoves; and he proposed that at least once a week a band should play in the Gardens, and that the public should be able to obtain the convenience of seats, as in the public gardens on the Continent. In the third number of the *Magazine* he began a series of articles on " Cottage Economy ;" and invited young architects to turn their thoughts to the erection of cottages, as well for labourers as for gardeners, which should be not only ornamental enough to please the gentlemen on whose grounds they were to be erected, but comfortable to those who were to live in them. These hints were followed up by many gentlemen : and I think I never saw Mr. Loudon more pleased than when a highly respectable gardener once told him that he was living in a new and most comfortable cottage, which his master had built for him; a noble marquess, who said that he should never have thought of it, but for the observations in Mr. Loudon's *Gardener's Magazine*, as they made him consider whether the cottage was comfortable or not, and that, as soon as he did so, he perceived its deficiencies. The fact is, that the greater part of the nobility and landed proprietors are, I believe, most anxious to make those around them as comfortable as possible, and only require their attention to be properly directed to the subject. In the fourth number of the *Gardener's Magazine* the subject of the reform of the Horticultural Society was resumed.; and it was continued in the succeeding numbers till 1830, when the desired result was at length effected.

Both in the early volumes of *The Gardener's Magazine*, and in the *Encyclopædia of Gardening*, Mr. Loudon had strongly advocated the necessity of having garden libraries; and in the second volume of *The Gardener's Magazine* he gave a list of books he considered suitable for a garden library, in which he included the *Encyclopædia of Plants* and the *Hortus Britannicus;* works then written, though they took so long in printing that they were not published till two or three years afterwards. It is very gratifying to find that numerous garden libraries were established in different parts of the country, in the course of two or three months after they were first suggested in *The Gardener's Magazine;* and that several letters appeared, from working gardeners, on the advantages and improvement which they had received from the books they thus obtained access to.

In the year 1827 Mr. Loudon suggested the idea of planting some public walk according to the natural system, and naming the trees in the way that has lately been done in Kensington Gardens. The same year the first notices were inserted of Horticultural Societies offering premiums for the production of certain vegetables, flowers, and fruits; a plan which has since been carried to a very great extent.

In the year 1828 *The Magazine of Natural History* was begun, being the first work of its kind; and this work, though not quite so successful as *The Gardener's Magazine,* was very popular, and has had numerous imitators. Towards the close of this year Mr. Loudon paid another visit to the Continent, to obtain information for a new edition of the *Encyclopædia of Agriculture.* After traversing France, he proceeded through Strasburg to Munich and Stuttgard; he afterwards visited Heidelberg and Carlsruhe, and returned by Metz to Paris, and thence to England. In *The Gardener's Magazine* for 1828 he began to give an account of this tour; and he continued it through several of the succeeding volumes, interspersing the descriptions of the various places he saw with a mass of valuable reflections on various subjects, which he conceived would be useful to gardeners. In the following year, 1829, he suggested the idea of having breathing zones, or unoccupied spaces half a mile broad, at different intervals around London; and in the next article to this he first suggested the idea of making use of the manure now carried to waste by the com-

mon sewers, a plan which has since engaged the attention of many talented persons, and which, probably,
will at no very distant period be carried into effect.
Another plan suggested by him about this period was for
establishing national schools, or, as he termed them, paro-
chial institutions for education.   In the same volume is a
suggestion for the establishment of a gardeners' fund for
the relief of the widows and families of deceased gardeners.

About this time Mr. Loudon formed his first acquaintance with me.   My father died in 1824; and, finding on
the winding up of his affairs that it would be necessary
for me to do something for my support, I had written a
strange wild novel called *The Mummy,* in which I had
laid the scene in the twenty-second century, and attempted
to predict the state of improvement to which this country
might possibly arrive.   Mr. Loudon chanced to see the
review of this book in the *Literary Gazette,* and, as among
other things I had mentioned a steam-plough, it attracted
his attention, and he procured the work from a circulating
library.   He read it, and was so much pleased with it,
that he published, in *The Gardener's Magazine* for 1828,
a notice of it under the head of " Hints for Improvements;" and he had from that time a great desire to become acquainted with the author, whom he supposed to
be a man.   In February, 1830, Mr. Loudon chanced to
mention this wish to a lady, a friend of his, who happened
to be acquainted with me, and who immediately invited
him to a party, where she promised him he should have
the wished-for introduction.   It may be easily supposed
that he was surprised to find the author of the book a
woman; but I believe that from that evening he formed
an attachment to me, and, in fact, we were married on the
14th of the following September.

Immediately after our marriage, Mr. Loudon began to
rewrite the *Encyclopædia of Gardening,* which was published in the course of the year 1831.   On the 1st of
October, 1830, he published the first part of a work, in
atlas folio, entitled *Illustrations of Landscape-Gardening*

and Garden Architecture; but, from the very expensive
nature of the work, and the limited number of subscribers,
he found it necessary to discontinue it, and it did not
proceed beyond the third part, which appeared in 1833.
In the beginning of the year 1831 he had an application to lay out a botanic garden at Birmingham, and
he agreed to do it merely on the payment of his expenses.
On this occasion I accompanied him; and, after spending
about six weeks in Birmingham, (which, though it is
my native town, I had not seen for several years,) we
made a tour through the North of England, visiting the
lakes in Cumberland and Westmoreland. It was at
Chester that we saw a copy of Mr. Paxton's *Horticultural
Register,* the first rival to *The Gardener's Magazine,* which
at the time we were married produced 750*l.* a year; but
which gradually decreased from the appearance of the *Horticultural Register,* till the period of Mr. Loudon's death,
immediately after which it was given up.

After visiting the beautiful scenery in Westmoreland
and Cumberland, we passed through Carlisle, and entered
Scotland by way of Longtown and Langholme. It happened that there was a fair at the latter place, and the
town was so exceedingly full that they not only could not
give us a bed, but we could not even find a place to sit
down. We had a four-wheeled phaeton with only one
horse, and, as we had travelled from Carlisle that day,
the animal was very much tired; it was also a serious annoyance to us, after having entered Scotland, to have to
return twenty miles into England, as we were told we
must do, Longtown being the nearest place where we
were likely to obtain accommodation for the night. Fortunately for us, Mr. Loudon, having heard that Mr. Bell,
who resided at Woodhouselee, only a few miles from Langholme, had a fine collection of American plants, determined to call there, and ask permission to see them. We
did so; and, when Mr. Bell heard how we were situated, he
most hospitably insisted on our staying at Woodhouselee
all night, though we were wholly strangers to him.

The next day we proceeded through Gretna Green and
Annan to Dumfries, in the neighbourhood of which we
staid about three weeks, spending part of the time at Close-
burn with Mr. Loudon's very kind friend Sir Charles
Menteath, and part at Jardine Hall with Sir William
and Lady Jardine.   We afterwards staid at Munches
and other seats in Dumfries-shire; and when we entered
Ayrshire, the county to which Mr. Loudon's family originally belonged, he was received with public dinners at
Ayr and Kilmarnock.   A public dinner was also pre-
paring for him at Glasgow; but while we were staying
at Crosslee Cottage, near Paisley, the residence of
Archibald Woodhouse, Esq., one of his most highly es-
teemed friends, he received a letter from Bayswater, in-
forming him of the severe illness of his mother, and her
earnest wish to see him.   Mr. Loudon was warmly attached to his mother, and as, unfortunately, we did not
receive the letter till late at night, for we had been dining
in the neighbourhood, we did not go to bed, but packed
up every thing so as to be able to set off with daylight
the next morning for Glasgow, where we left Mr. Loudon's
man with the horse and carriage, and proceeded to Edinburgh by coach, though we could only get outside places,
and it rained; besides which, Mr. Loudon had never ridden
on the outside of a coach since his knee had become stiff,
and he could not ascend the ladder without the greatest
difficulty.   Nothing, however, could stop him in the per-
formance of what he considered his duty, and indeed I
believe his eagerness to see his mother overpowered every
other feeling,   It was also a singular circumstance, that,
on his return to Edinburgh after an absence of nearly
thirty years, he should be obliged to pass through it
almost without stopping; yet such was the case, as we
found on our arrival at the inn that a packet was just
about to sail for London, and that if we did not avail
ourselves of it we should be compelled to wait several
days.   We, therefore, hurried down to the pier; and, finding that the captain of the vessel was just going on board,

we hired a boat, and were luckily in time to save our passage. We had a very quick voyage, and arrived at Bayswater about half an hour after the letter we had sent from Glasgow to announce that we were coming. Mr. Loudon's mother was so delighted to see her son, that she seemed partially to revive; so much, indeed, that we had hopes of her recovery. Nature, however, was too far exhausted, and she died about six weeks after our return, in October, 1831.

In 1832 Mr. Loudon commenced his *Encyclopædia of Cottage, Farm, and Villa Architecture*, which was the first work he ever published on his own account; and in which I was his sole amanuensis, though he had several draughtsmen. The labour that attended this work was immense; and for several months he and I used to sit up the greater part of every night, never having more than four hours' sleep, and drinking strong coffee to keep ourselves awake. The *First Additional Supplement* to the *Hortus Britannicus* was also prepared and published in 1832.

The great success of the *Cottage Architecture*, which is perhaps the best and most useful of all Mr. Loudon's works, tempted him to publish the *Arboretum Britannicum* also on his own account. He had long intended to write a work on the hardy trees of Great Britain; but he did not contemplate the expenses which he should incur by so doing. When, however, the *Arboretum* was once begun, he found it was impossible to compress it into the limits originally intended; and, in his determination to make the work as perfect as possible, he involved himself in the difficulties which hastened his death. Notwithstanding the immense labour attending the *Arboretum*, which was published in monthly numbers, Mr. Loudon, in March, 1834, began *The Architectural Magazine*, the first periodical devoted exclusively to architecture; though, like *The Magazine of Natural History* and *The Gardener's Magazine*, it only served as a pioneer to clear the way for others, which afterwards followed in the same course with much greater success.

From the year 1833 to Midsummer 1838 Mr. Loudon underwent the most extraordinary exertions both of mind and body. Having resolved that all the drawings of trees for the *Arboretum* should be made from nature, he had seven artists constantly employed, and he was frequently in the open air with them from his breakfast at seven in the morning till he came home to dinner at eight in the evening, having remained the whole of that time without taking the slightest refreshment, and generally without even sitting down. After dinner he resumed the literary part of the work, and continued writing, with me as his amanuensis, till two or three o'clock in the morning. His constitution was naturally very strong; but it was impossible for any human powers to bear for any lengthened period the fatigue he underwent. In 1836 he began *The Suburban Gardener,* which was also published in monthly numbers, so that he had five monthly works going on at the same time. He soon found, however, that three monthly works, besides the *Arboretum,* were as much as his health would permit him to undertake the management of, and he disposed of *The Magazine of Natural History* to Mr. Charlesworth. In 1838 he also gave up *The Architectural Magazine,* and at Midsummer in that year he finished the *Arboretum Britannicum.* He was now in circumstances that would have discouraged almost any person but himself. His health was very seriously injured, partly by what was supposed to be a liver complaint, and partly by an enormous swelling in his right knee, which some of the most eminent medical men in London supposed to be produced by a disease in the bone. In addition to the large sums in ready money he had paid to the artists and other persons employed during the progress of the *Arboretum,* he found at its conclusion that he owed ten thousand pounds to the printer, the stationer, and the wood-engraver who had been employed on that work. His creditors, however, did not press him for their money, but gave him a chance of reaping the benefit of his labours at some future time, by consenting to wait till they were paid by the sale of the

Arboretum and the *Cottage Architecture,* upon condition
that he placed these works in the hands of Messrs. Longman, to hold for the creditors till the debt was paid.

Notwithstanding the state of his knee, which was now
such that he was unable to walk without assistance, immediately on the completion of the *Arboretum* he arranged and published his *Hortus Lignosus Londinensis ;*
and in the last number of *The Suburban Gardener,* which
was finished about this time, he informed the public that
he intended to resume his profession of landscape-gardener, and that he would not only go out, but give advice
at home, on any plans that might be sent to him. To us,
who saw the state of his health, this intimation gave the
greatest pain, and we determined to do every thing in our
power to prevent the necessity of his exerting himself.
Two of his sisters learned wood-engraving ; and I, having
acquired some knowledge of plants and gardens during
the eight years I had acted as his amanuensis, began to
write books on those subjects myself. In the mean time,
he grew so much worse, that we had very little hope of
his recovery, till he placed himself under the care of
William Lawrence, Esq. ; when that eminent surgeon
took a different view of the case from what had been before
entertained, and by his mode of treatment rapidly restored
him to health.

In 1839 Mr. Loudon began to lay out the Arboretum
so nobly presented by the late Joseph Strutt, Esq., to the
town of Derby. In the same year he published his edition
of *Repton,* and his *Second Additional Supplement to the
Hortus Britannicus.* In 1840 he accepted the editorship of
The Gardener's Gazette, which, however, he only retained
about a year.

In 1840, Mr. Loudon, having a great desire to examine
some of the trees in the Jardin des Plantes, in order
to identify some of the species of *Cratæ`gus,* went to
Paris ; and, as his health was beginning again to decline,
I went with him, taking with me our little daughter
Agnes, who, from this time, was always the companion of

our journeys.   We went by way of Brighton, Dieppe, and
Rouen, to Paris, ascending the Seine; and we remained
in France about two months.

When Mr. Loudon left Scotland so abruptly in 1831, he
promised his friends to return the following year, and, indeed, fully intended to do so; but various circumstances
occurred to prevent him, and it was not till 1841 that
he was able to fulfil his engagement.   In the summer
of that year, however, soon after the publication of the
Supplement to the Encyclopædia of Plants, Mr. Loudon,
Agnes, and myself, went from London to Derby, and,
after spending a few days with our kind and excellent
friend Mr. Strutt, we proceeded through Leeds to Manchester.   It rained heavily when we arrived at Leeds;
but, Mr. Loudon having determined to visit the Botanic
Garden, we went there in a most awful thunder-storm,
and the whole of the time we were in the garden the
rain descended in torrents.   We were all wet, and we had
no time to change our clothes, as, on our return to the
station, we found the last train to Manchester ready to
start, and Mr. Loudon was most anxious to proceed thither
without delay.   When we arrived at Manchester, he
was far from well; but notwithstanding, the next morning,
though it still rained heavily, he insisted upon going
to the Botanic Garden.   Here he increased his cold, and
when we returned to the inn he was obliged to go to bed.
The next morning, however, he would go on to Liverpool;
and, though he was so ill there that when we drove to the
Botanic Garden he was unable to get out of the coach,
and was obliged to send me to look at some plants he
wished to have examined, he would sail for Scotland that
night.   He was very ill during the voyage, and when we
landed at Greenock he was in a high fever.   He persisted,
however, in going by the railway to Paisley, and thence to
Crosslee Cottage, where we had promised to spend a few
days with our kind friends Mr. and Mrs. Woodhouse.
When we arrived there, however, he was obliged instantly
to go to bed.   A doctor was sent for, who pronounced

his disease to be a bilious fever, and for some time his life appeared in great danger.

It was six weeks before he could leave his bed; but as soon as he was able to sit up he became anxious to resume his labours; and, taking leave of our kind friends, we set out on a tour through the South of Scotland, visiting every garden of consequence on our route, and making notes of all we saw. Notwithstanding all he had suffered during his severe illness, and the state of weakness to which he was reduced, he exerted himself to see every thing; and he was never deterred, either by fatigue or wet weather, from visiting every garden that he heard contained any thing interesting. After travelling about a fortnight we reached Edinburgh, but Mr. Loudon only staid one night; and, leaving Agnes and me there, he proceeded on the 13th of August alone to Glasgow, on his road to Stranraer, where he was going to lay out the grounds at Castle Kennedy, for the Earl of Stair.

On the 1st of September he returned to Edinburgh, which of course he found greatly changed since he had resided there thirty-seven years before; and for the next fortnight he had great pleasure in showing me the places he had known when a boy. On the 13th of September, having hired a carriage at Edinburgh, we set out on our return home by land; and at Newcastle we spent two or three days with our friends Mr. and Mrs. Sopwith, where Mr. Loudon was highly gratified with the arrangement of Mr. Sopwith's library, which we found a perfect temple of order.

On leaving Newcastle we travelled through Chester-le-Street to Durham, visiting nearly all the fine places in that county, particularly Raby Castle; and afterwards we proceeded to Darlington, where we took the railroad to York. We stayed three or four days in this city, and then we returned to London by the railroad.

In December, 1841, appeared the first number of the *Encyclopædia of Trees and Shrubs*, the work consisting of ten monthly numbers. The abridgement of the *Hortus Lignosus Londinensis* was published immediately on the conclusion of the *Encyclopædia of Trees and Shrubs*; and in May, 1842, appeared the *First Additional Supplement to the Encyclopædia of Cottage Architecture.*

In addition to the works which have been enumerated, Mr. Loudon contributed to several others, such as the *Encyclopædia of Domestic Economy,* and *Brande's Dictionary of Science, Literature, and Art.* He also wrote the article Planting for the new edition of the *Encyclopædia Britannica.*

Early in March, 1842, he had an attack of inflammation of the lungs, and, on his recovery, we went down to Brighton for some weeks. We afterwards made a tour through Somersetshire, Devonshire, and part of Cornwall; and, on our return to Exeter, Mr. Loudon went to Barnstaple, in the neighbourhood of which he was about to lay out some grounds for Lord Clinton, sending Agnes and myself back to London. When he returned home, I noticed that he had a slight cough; but, as it was trifling, it did not make me uneasy, particularly as his spirits were good. He now finished his *Suburban Horticulturist,* which had been begun two years before, but had been stopped on account of his illness in Scotland; and this work was published by Mr. Smith of Fleet Street, all his other works, from the appearance of the *Encyclopædia of Gardening,* having been published by Messrs. Longman.

In 1843 his time was chiefly occupied by his work on *Cemeteries,* with which he took extraordinary pains, and which was very expensive from the number of the engravings. In August we were invited to Derby to pay another visit to Mr. Strutt, but he was too ill to go, and the doctors pronounced his complaint to be a second attack of inflammation of the lungs.

Previously to Mr. Loudon's illness, I had agreed to write a little book on the Isle of Wight, and to visit it for this purpose. This arrangement I now wished to give up; but his medical men advised us to go, as they thought the air of the Isle of Wight might reestablish his health. Strange to say, up to the time of our leaving home I had no idea that his illness was at all dangerous; but the fact was, I had seen him recover so often when every one thought he was dying, that I had become accustomed to place little reliance on what was said of his attacks by others. When we reached the Isle of Wight, however, I was struck with a degree of listlessness and want of energy about him that I had never seen before. He became rapidly worse while we were in the island, and most eager to leave it. On our arrival at Southampton, where he was laying out a cemetery, he felt better; and, taking a lodging there, he sent Agnes and myself back to town. In a fortnight I went down to see him, and I shall never forget the change I found in him. The first look told me he was dying. His energy of mind had now returned. He not only attended to the laying out of the cemetery at Southampton; but during his stay in that town he corrected the proofs of the second *Supplement* to his *Encyclopædia of Agriculture,* and then went alone to Bath, in spite of my earnest entreaties to be permitted to accompany him. At Bath he inspected the ground for another cemetery, and also the grounds of a gentleman named Pinder, though he was obliged to be wheeled about in a Bath chair. He then went, still alone, to Kiddington, the seat of Mortimer Ricardo, Esq., near Enstone, in Oxfordshire; where he was also obliged to be wheeled round the grounds in a chair. When about to leave Kiddington he appeared so ill, that Mr. Ricardo offered to send a servant with him to town.

He returned to Bayswater on the 30th of September, 1843, and at last consented to call in medical aid, though he was by no means aware of his dangerous state. He supposed, indeed, that the pain he felt, which was on the right side, proceeded from an affection of the liver; as both times, when he had inflammation of the lungs, the pain was on the left side. On the 2d of October I went with him to call on Mr. Lawrence, in whom he had the greatest confidence; and that gentleman told him without hesitation that his disease was in his lungs. He was evidently very much struck at this announcement, but, as he had the fullest reliance on Mr. Lawrence's judgment, he was in-

stantly convinced that he was right; and, I think, from
that moment he had no hope of his ultimate recovery,
though, in compliance with the wishes of different friends,
he afterwards consulted several other eminent medical men,
of whom Dr. Chambers and Mr. Richardson attended him
to the last.

As soon as Mr. Loudon found that his disease was
likely to prove fatal, he determined, if possible, to finish
the works he had in hand, and he laboured almost night
and day to do so. He first, with the assistance of his
draughtsman, finished a plan for Baron Rothschild; then
one for Mr. Ricardo, another for Mr. Pinder, and, finally,
a plan for the cemetery at Bath. He had also engaged
to make some additional alterations in the grounds of
Mr. Fuller at Streatham, and he went there on the 11th
of October, but he was unable to go into the garden; and
this was the last time he ever attempted to visit any place
professionally. He continued, however, to walk in the
open air in his own garden, and in the grounds of Mr.
Hopgood, nurseryman, at Craven Hill, for two or three
days longer, though his strength was fast decreasing; and
after the 16th of October he did not leave the house, but
confined himself to his bedroom and a drawingroom on the
same floor. Nothing could be more awful than to watch
him during the few weeks that yet remained of his life.
His body was rapidly wasting away; but his mind remained in all its vigour, and he scarcely allowed himself
any rest in his eagerness to complete the works that he had
in hand. He was particularly anxious to finish his *Self-
Instruction for Young Gardeners*, which is published nearly
in the state he left it, though had he lived it would probably have been carried to a much greater extent. About
the middle of November, the medical men who attended
my poor husband pronounced his disease to have become
chronic bronchitis; and this information, combined with
the pressure of pecuniary difficulties, had a powerful
effect upon him. He now made an effort that can only
be estimated by those who know the natural independence of his mind, and the pain it gave him to ask even

a trifling favour. He wrote a letter stating his situation, and that the sale of 350 copies of the *Arboretum* would free him from all his embarrassments. This letter he had lithographed, and he sent copies of it to all the nobility who took an interest in gardening. The result was most gratifying. The letter was only dated the 1st of December, and he died on the 14th of that month; and yet in that short space of time the noblemen he appealed to, with that kindness which always distinguishes the English aristocracy, purchased books to the amount of 360*l.* Mr. Loudon had intended to forward similar letters to all the landed proprietors and capitalists; and, though only a few were sent, they were responded to with equal kindness. Our munificent and noble-minded friend Joseph Strutt, Esq., took ten copies; and letters from two of our kindest friends (William Spence, Esq., and Robert Chambers, Esq.), ordering copies of the *Arboretum*, arrived the very day he died.

This appeal was principally rendered necessary by the pecuniary difficulties I have alluded to, and which, undoubtedly, hastened his death. The debt on the *Arboretum*, which, as already stated, was originally 10,000*l.*, had, by the sale of that book and of the *Cottage Architecture*, been reduced to 2400*l.*; but he had incurred an additional debt of 1200*l.* by publishing the *Encyclopædia of Trees and Shrubs*, his edition of *Repton*, and other works, on his own account, though all his creditors agreed to the same terms, viz. to wait for their money until they were paid by the sale of the works themselves, on condition of Messrs. Longman holding the stock of books in trust, and not paying any of the proceeds of the works to Mr. Loudon till the demands of his creditors were fully satisfied. Unfortunately, however, one of the creditors, the engraver, became a bankrupt, and his assignees began to harass Mr. Loudon for the debt due to them, which was about 1500*l.*, threatening to make him a bankrupt, to arrest him for the sum, &c. I believe they could not have carried their threats into execution without the consent of Mr. Spottiswoode, and Messrs. Smith and

Chapman, who were the other creditors, and who behaved most kindly and honourably throughout. But the agitation attendant on the numerous letters and consultations respecting this affair proved fatal to my poor husband.

On Wednesday the 13th of December, 1843, he sent me into London to see the assignees, and to endeavour to bring them to terms, our kind and excellent friend, the late Mr. Joseph Strutt, having promised to lend us money for that purpose. The assignees, however, refused to accept the terms we offered, unless Mr. Loudon would also give up to them his edition of *Repton*, which he was most unwilling to do, as the debt on that work was comparatively small; and, consequently, he had reason to hope that the income produced by it would be soonest available for the support of his family. He was accordingly very much agitated when I told him the result of my mission; but he did not on that account relax in his exertions; on the contrary, he continued dictating *Self-Instruction* till twelve o'clock at night. When he went to bed he could not sleep, and the next morning he rose before it was light. He then told me he had determined to sacrifice his edition of *Repton* in order to have his affairs settled before he died; adding " but it will break my heart to do so." He repeated, however, that he would make the sacrifice, but he seemed reluctant to send me into town to give his consent; and most fortunate was it, as, if I had gone to town that morning, I should not have been with him when he died. He now appeared very ill, and told me he thought he should never live to finish *Self-Instruction;* but that he would ask his friend Dr. Jamieson, to whom he had previously spoken on the subject, to finish the work for him. Soon after this he became very restless, and walked several times from the drawingroom to his bedroom and back again. I feel that I cannot continue these melancholy details: it is sufficient to say, that, though his body became weaker every moment, his mind retained all its vigour to the last, and that he died standing on his feet. Fortunately, I perceived a change taking place in his countenance, and I had just time to clasp my arms round him, to save him from falling, when his head sank upon my shoulder, and he was no more.

I do not attempt to give any description of the talents or character of my late husband as an author; his works are before the world, and by them he will be judged; but I trust I may be excused for adding, that in his private capacity he was equally estimable as a husband and a father, and as a master and a friend. He was also a most dutiful son and most affectionate brother.

It was on the anniversary of the death of Washington (the 14th of December) that Mr. Loudon died, and he was buried, on the 21st of December, in the cemetery at Kensall Green. When the coffin was lowered into the grave, a stranger stepped forward from the crowd and threw in a few strips of ivy. This person, I was afterwards informed, was an artificial flower maker, who felt grateful to Mr. Loudon for having given him, though a stranger, tickets for admission to the Horticultural Gardens, and who, never having been able to thank Mr. Loudon in person, took this means of paying a tribute to his memory.

In addition to the preceding memoir, I have ventured to reprint the following anecdotes, which originally appeared in the *Derby Reporter*; as they were written shortly after Mr. Loudon's death by a young man who knew him well, having acted as his draughtsman for upwards of nine years.

" Mr. Loudon's love of truth, like that of every great and good man, was perfect; and he would at all times make any personal sacrifice for its cause, or to punish falsehood. An instance occurred in 1831, which, though trifling, showed his strictness in this matter. He had a young man, an amanuensis, who had been with him for some years, and of whom he was exceedingly fond. He sent this person one morning from Bayswater to make a tracing at the residence of the celebrated Mr. Telford, Westminster. The youth being delighted at getting out from the confinement of an office, forgot, until he arrived at the place, that he had no pencils or tracing-paper with him, and unfortunately did not think of purchasing them. He thought he should look so foolish to return and say he had forgotten the materials, so he made up his mind to tell Mr. Loudon an untruth for the first time. He returned, and said falteringly, ' Mr. Telford was not at home.' Mr. Loudon fixed his keen eye upon him, and observed, ' Did I understand you to say that Mr. Telford was not at home?' The answer was in the affirmative. ' Very well,' said Mr. Loudon; and shortly after rang for the man-servant, and ordered the phaeton. He drove direct to Westminster, and found that Mr. Telford had been confined to his house for some weeks, unwell. He returned, paid the amanuensis his salary, gave him something extra to pay his lodgings for a week, and immediately discharged him; remarking that, however valuable his services were, he (Mr. L.) would not suffer any one to remain a single night longer in his house who had told him a falsehood.

c

" Mr. Loudon also mixed with his love of truth, determination. About this time, an officer, rather a public
character, was in the habit of visiting the family. His
ungentlemanly manners, one day, gave Mr. Loudon of-
fence, and he determined not to see him again. About
the gentleman's usual time of coming, when the bell was
rung, Mr. Loudon told the servant that if *that* were ——,
' just tell him I cannot see him.' ' Shall I say that you
are not at home, sir?' said the servant. ' No,' was
Mr. Loudon's reply ; ' you would then tell a falsehood,
which you must not do. Just tell the gentleman I cannot
see him.'

" His love of order was also very great. The books in
the library, and manuscripts in his study, were so arranged
that he could at any time put his hand upon any book or
paper that he might want, even in the dark. He instilled
this system of order into the minds of his clerks too ; for,
when any new one came, his invariable instructions were —
' Put every thing away before you leave at night, as if
you never intended to return.'

" He was also a man of great punctuality as to time,
money matters, and in every other respect. When any of
his clerks happened to be behind time in the morning, he
would take no notice for a few times ; but, if it were often
repeated, he would say very quietly but sarcastically —
' Oh, if 9 o'clock is too early for you, you had better
come at 11 or 12 ; but let there just be a fixed hour, that
I may depend upon you.'

" Mr. Loudon was a man of great fortitude and unwearied industry. The morning that Doctors Thompson
and Lauder called upon him for the purpose of amputating
his right arm, they met him in the garden, and asked if
he had fully made up his mind to undergo the operation.
' Oh, yes, certainly,' he said; ' it was for that purpose
I sent for you;' and added very coolly, ' but you had
better step in, and just have a little lunch first before you
begin.' After lunch he walked up stairs quite com-
posedly, talking to the doctors on general subjects. When

all the ligatures were tied, and every thing complete, he was about to step down stairs, as a matter of course, to go on with his business; and the doctors had great difficulty to prevail upon him to go to bed.

" As a man of industry, he was not surpassed by any one. Deducting for the time he has been poorly, he has, during three fourths of his literary career, dictated about five and a half printed octavo pages of matter every day on an average. He has been frequently known to dictate to two amanuenses at the same time. He often used to work until 11 and 12 o'clock at night, and sometimes all night. It may not be amiss to mention here, as illustrative of his love of labour, that, whilst his man-servant was dressing him for church on the day of his marriage, he was actually dictating to his amanuensis the whole time.

" Although Mr. Loudon was a matter-of-fact man, he had nevertheless a good deal of poetry in his soul. The writer happened to dine with him the day that he attended Dr. Southwood Smith's Anatomical Lecture on the body of his friend Jeremy Bentham. Just at the moment the lecturer withdrew the covering from the face of the corpse the lightning flashed, and an awful burst of thunder pealed forth —

' Crush'd horrible, convulsing heaven and earth!'

Mr. Loudon, during dinner, gave a most touching, poetical, and graphic description of the lecture, and the circumstances attending it; and every one present could see how deeply he felt the loss of his friend Bentham.

" Mr. Loudon was a man, like most good men, rather easily imposed upon. He, contrary to the ways of the world, looked upon every man as a good man until he had proved him otherwise; but when he had done so, he was firm in his purpose. He was a warm friend, an excellent husband, an amiable brother, and a most affectionate and dutiful son. Altogether

' He was a man, take him for all in all,
We shall not look upon his like again.'

"ELEGY.

"HARK! hark! the sound — 'tis a fun'ral knell
    Borne on the breath of day —
The mournful voice of the deep-toned bell —
    For a spirit has wing'd his way.

" 'Tis not the man of wealth and state
    That the world has now to mourn;
'Tis not the man that gold makes great
    Who now to the tomb is borne.

" No! no! we grieve, in the friend now gone,
    No flattering slave of state;
But the world has lost by the death of one
    Whose mind was truly great.

" He wielded no sword in his country's cause,
    But his pen was never still;
He studied each form of Nature's laws,
    To lessen each human ill.

" That voice is hush'd! — and lost the sound
    Employ'd to raise the poor;
But the echo shall, by his works, be found
    To reach the rich man's door.

" He wakes no more! — for the sleep of death
    Encircles the earthly frame;
But the mind — so strong while it dwelt on earth —
    Secured a living fame.

" His pen is still! — and his spirit fled
    To brighten a world on high:
The cold, cold earth is his lowly bed;
    But his *name* shall never die!

J. R."

" *Chatsworth.*

SELF-INSTRUCTION

FOR

YOUNG GARDENERS, FORESTERS,

&c. &c.

INTRODUCTION.

THE young men for whom this book is intended are chiefly such as have received but a very imperfect rudimentary education, or have forgotten, in a great measure, what they have been taught. Though we have chiefly had in view young gardeners, yet we have had an eye also to the other classes mentioned in the titlepage; partly for the sake of those classes, and partly because gardeners are frequently required to practise the business of forester, and sometimes that of bailiff; occasionally they are elevated to the stewardship of an estate; and it is no uncommon thing for a gardener, after he has made a little money, to retire from servitude, and become a nurseryman or a rent-paying farmer.

We have here chiefly confined ourselves to such studies as concern horticulture and agriculture as mechanical arts; those which relate to them as arts of culture, such as the Geography of Natural History, Geology, Meteorology, Chemistry, and Physiology, require to be studied in separate works. These sciences will be found treated on at sufficient length in the introductory parts of the *Encyclopædias of Gardening and Agriculture*, and in the *Suburban Horticulturist*.

B

CHAPTER I.

ARITHMETIC.

ARITHMETIC is the science of numbers, or the art of reckoning or calculating by them. Unlike many other arts, it is not only in use in civilised society, but among barbarians; and not only among grown-up persons, but among children as soon as they can speak. The only change that can take place in numbers is by adding to them, or diminishing them; and hence the whole art of arithmetic may be comprehended in addition and subtraction. Multiplication is only addition several times repeated, and division a series of subtractions.

We shall suppose the reader to be conversant with the four common rules of arithmetic; and, therefore, we shall proceed at once to what are called Vulgar and Decimal Fractions.

SECTION I.

VULGAR FRACTIONS.

IN order to understand the nature of vulgar fractions, we must suppose unity (or the number 1) divided into several equal parts. One or more of these parts is called a *fraction*, and is represented by placing one number in a small character above a line, and another under it. For example, two-fifth parts is written thus $\frac{2}{5}$. The number under the line (5) shows how many parts unity is divided into, and is called the *denominator*. The number above the line (2) shows how many of these parts are represented, and is called the *numerator*.

It follows from the manner of representing fractions, that when the numerator is increased, the value of the fraction becomes greater; but when the denominator is increased, the value becomes less. Hence we may infer, that if the numerator and denominator be both increased, or both diminished, in the same proportion, the value is not altered; and therefore, if we multiply both by any number whatever, or divide them both by any number of which both are measures, we shall obtain other fractions of equal value. Thus, every fraction may be expressed in a variety of forms, which have all the same signification.

A fraction annexed to an integer or whole number makes a mixed number; for example, five and two third parts, or $5\frac{2}{3}$. A fraction whose numerator is greater than its denominator is called an *improper fraction;* for example, seventeen third parts, or $\frac{17}{3}$. Fractions of this kind are greater than unity. Mixed numbers may be represented in the form of improper fractions, and improper fractions may be reduced to mixed numbers, and sometimes to integers. As fractions, whether proper or improper, may be represented in different forms, we must explain the method of reducing them from one form to another before we consider the other operations.

PROBLEM I. To reduce mixed Numbers to improper Fractions.

Rule. *Multiply the integer by the denominator of the fraction, and to the product add the numerator. The sum is the numerator of the improper fraction sought, and is placed above the given denominator.*

$$\text{Ex. } 5\frac{2}{3}=\frac{17}{3}=\frac{5\times 3+2}{3}.$$

Here 5 the integer is multiplied by
  3 the denominator of the fraction, and
to 15, the product, we add
  2 the numerator given. Hence
17 is the numerator sought: beneath which we write the denominator 3, and thus obtain the improper fraction $\frac{17}{3}$.

Or thus. Because one is equal to two halves, or 3 third parts, or 4 quarters, and every integer is equal to twice as many halves, or four times as many quarters, and so on, therefore every integer may be expressed in the form of an improper fraction, having an assigned denominator. The numerator is obtained by multiplying the integer into the denominator. Hence the reason of the foregoing rule is evident. 5 reduced to an improper fraction, whose denominator is 3, makes $\frac{15}{3}$; and this added to $\frac{2}{3}$ amounts to $\frac{17}{3}$.

PROBLEM II. To reduce improper Fractions to whole or mixed Numbers.

Rule. *Divide the numerator by the denominator.*

$$\text{Ex.} \quad \tfrac{112}{17} = 112 \div 17 = 6\tfrac{10}{17}.$$

$$\text{Or thus,} \quad 17) 112 (6\tfrac{10}{17}$$
$$\underline{102}$$
$$10$$

This problem is the converse of the former, and needs no further illustration.

PROBLEM III. To reduce Fractions to Lower Terms.

Rule. *Divide both numerator and denominator by any number which measures both, and place the quotients in the form of a fraction.*

$$\text{Ex.} \ \tfrac{135}{360} = \tfrac{27}{72} = \tfrac{3}{8}; \ \text{or} \ \tfrac{135}{360} \div 45 = \tfrac{3}{8}.$$

Here we observe that 135 and 360 are both measured by 5, and the quotients form $\frac{27}{72}$, which is a fraction of the same value as $\frac{135}{360}$ in lower terms. Again; 27 and 72 are both measured by 9, and the quotients form $\frac{3}{8}$, which is still of equal value, but in lower terms.

It is generally sufficient, in practice, to divide by such measures as are found to answer on inspection. But if it be required to reduce a fraction to the lowest possible terms, we must divide the numerator and denominator by the greatest number which measures both. What number this is may not be obvious, but will always be found by the following rule.

To find the greatest common measure of two numbers, divide the greater by the less and the divisor by the remainder continually till nothing remains; the last divisor is the greatest common measure.

Ex. Required the greatest common measure, or the number which measures 475 and 589.

$$475) 589 (1$$
$$475$$
$$\overline{}$$
$$114) 475 (4$$
$$456$$
$$\overline{}$$
$$19) 114 (6$$
$$114$$
$$\overline{}$$

Here we divide 589 by 475, and the remainder is 114; then we divide 475 by 114, and the remainder is 19; then we divide 114 by 19, and there is no remainder; from which we infer that 19, the last divisor, is the greatest common measure.

To explain the reason of this we must observe, that any number which measures two others will also measure their sum and their difference, and will measure any multiple of either. In the foregoing example, any number which measures 589 and 475 will measure their difference, 114, and will measure 456, which is a multiple of 114; and any number which measures 475 and 456 will also measure their difference, 19. Consequently, no number greater than 19 can measure 589 and 475. Again; 19 will measure them both, for it measures 114, and therefore measures 456, which is a multiple of 114, and 475, which is just 19 more than 456; and because it measures 475 and 114, it will measure their sum, 589. To reduce $\frac{475}{589}$ to the lowest possible terms, we divide both numbers by 19, and it comes to $\frac{25}{31}$.

If there be no common measure greater than 1, the fraction is already in the lowest terms.

If the greatest common measure of 3 numbers be required, we find the greatest measure of the two first, and then the greatest measure of that number and the third. If there be more numbers, we proceed in the same manner.

PROBLEM IV.   To reduce Fractions to others of equal Value that have the same Denominator.

Rule. 1st. *Multiply the numerator of each fraction by all the denominators except its own; the products are numerators to the respective fractions sought.*   2d. *Multiply all the denominators into each other; the product is the common denominator.*

Ex. $\frac{4}{5}$ and $\frac{7}{9}$ and $\frac{3}{8} = \frac{288}{360}$ and $\frac{280}{360}$ and $\frac{135}{360}$; for
$4 \times 9 \times 8 = 288$ first numerator.
$7 \times 5 \times 8 = 280$ second numerator.
$3 \times 5 \times 9 = 135$ third numerator.
$5 \times 9 \times 8 = 360$ common denominator.

Here we multiply 4, the numerator of the first fraction, by 9 and 8, the denominators of the two others; and the product, 288, is the numerator of the fraction sought, equivalent to the first.   The other numerators are found in like manner, and the common denominator, 360, is obtained by multiplying the given denominators 5, 9, 8, into each other.   In the course of the whole operation, the numerators and denominators of each fraction are multiplied by the same number, and therefore their value is not altered.

The fractions thus obtained may be reduced to lower terms, if the several numerators and denominators have a common measure greater than unity.   Or, after arranging the number for multiplication, as is done above if the same number occur in each rank, we may strike them out and neglect them; and if numbers which have a common measure occur in each, we may dash them out and use the quotients in their stead; or any number, which is a multiple of all the given denominators, may be used as a common denominator. Sometimes a number of this kind will occur on inspection, and the new numerators are found by multiplying the given ones by the common denominator, and dividing the products by the respective given denominators.

If the articles given for any operation be mixed numbers, they are reduced to improper fractions by Problem I. If the answer obtained be an improper fraction, it is reduced to a mixed number by Problem II. And it is convenient to reduce fractions to lower terms, when it can be done, by Problem III., which makes their value better apprehended, and facilitates any following operation. The reduction of fractions to the same denominator by Problem IV. is necessary to prepare them for addition or subtraction, but not for multiplication or division.

———

Subsect. I. — ADDITION OF VULGAR FRACTIONS.

Rule. *Reduce them, if necessary, to a common denominator; add the numerators, and place the sum above the denominator.*

Ex. 1. $\frac{3}{5} + \frac{2}{9} = \frac{27}{45} + \frac{10}{45}$, by Problem IV. $= \frac{37}{45}$.

Ex. 2. $\frac{5}{7} + \frac{8}{9} + \frac{9}{10} = \frac{450}{630} + \frac{560}{630} + \frac{567}{630} = \frac{1577}{630}$, by Problem II. $= 2\frac{317}{630}$.

The numerators of fractions that have the same denominator signify like parts; and the reason for adding them is equally obvious, as that for adding shillings or any other inferior denomination.

Mixed numbers may be added by annexing the sum of the fractions to the sum of the integers. If the former be a mixed number, its integer is added to the other integers.

Subsect. II. — SUBTRACTION OF VULGAR FRACTIONS.

Rule. *Reduce the fractions to a common denominator; subtract the numerator of the subtrahend from the numerator of the minimend, and place the remainder above the denominator.*

Ex. Subtract $\frac{2}{7}$ from $\frac{5}{12}$, and the remainder will be $\frac{11}{84}$.

For $\left\{ \begin{array}{l} \frac{5}{12} = \frac{35}{84} \\ \frac{2}{7} = \frac{24}{84} \end{array} \right\}$ Minimend by Problem IV. Subtrahend. For if from 35 we take 24 the remainder is $\overline{11}$;

hence the difference of the fractions $\frac{5}{12}$ and $\frac{2}{7}$ is $\frac{11}{84}$.

Divide the same line into 12, 7, and 84 equal parts; then through $\frac{5}{12}$ and $\frac{2}{7}$ draw lines at right angles to the original and parallel lines (*fig.* 1.), and you will cut off $\frac{11}{84}$ths of the third line, which mechanically demonstrates the intellectual operation by figures.

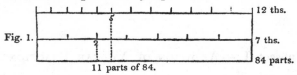

Fig. 1.   12 ths.
            7 ths.
            84 parts.
    11 parts of 84.

To subtract a fraction from an integer. *Subtract the numerator of the fraction from the denominator, place the remainder above the denominator, and prefix to this the integer diminished by unity.*

Ex. Subtract $\frac{3}{5}$ from 12 : the remainder $11\frac{2}{5}$; for $\frac{12}{1} - \frac{3}{5} = \frac{60-3}{5} = \frac{57}{5} = 11\frac{2}{5}$.

Subsect. III. — MULTIPLICATION OF VULGAR FRACTIONS.

Rule. *Multiply the numerators of the factors together for the numerator of the product, and the denominators together for the denominator of the product.*

Ex. 1. $\frac{2}{3} \times \frac{5}{7} = \frac{10}{21}$.
$2 \times 5 = 10$ num.
$3 \times 7 = 21$ den.

Ex. 2. $8\frac{2}{5} \times 7\frac{3}{4} = \frac{1302}{20} = 65\frac{2}{20}$.
$8\frac{2}{5} = \frac{42}{5}$ by Prob. I.
$7\frac{3}{4} = \frac{31}{4}$ by ditto.
$42 \times 31 = 1302$.
$5 \times 4 = 20$.

To multiply $\frac{5}{7}$ by $\frac{2}{3}$, is the same as to find what two third parts of $\frac{5}{7}$ comes to. If one third part only had been required, it would have been obtained by multiplying the denominator 7 by 3, because the value of fractions is lessened when their denominators are increased: and this comes to $\frac{5}{21}$; and, because two thirds were required, we must double that fraction, which is done by multiplying the numerator by 2, and comes to $\frac{10}{21}$. Hence we infer that fractions of fractions, or compound fractions, such as $\frac{2}{7}$ of $\frac{3}{5}$, are reduced to simple ones by multiplication. The same method is followed when the compound fraction is expressed in three parts or more.

Subsect. IV. — DIVISION OF VULGAR FRACTIONS.

Rule I. *Multiply the numerator of the dividend by the denominator of the divisor. The product is the numerator of the quotient.*

II. *Multiply the denominator of the dividend by the numerator of the divisor. The product is the denominator of the quotient.*

Ex. Divide $\frac{2}{5}$ by $\frac{7}{9}$, and the quotient is $\frac{18}{35}$.

For $\begin{cases} 2 \times 9 = 18 \\ 5 \times 7 = 35 \end{cases}$ or invert the divisor, and proceed as in multiplication.

To explain the reason of this operation, let us suppose it required to divide $\frac{2}{5}$ by 7, or take one seventh part of that fraction. This is obtained by multiplying the denominator by 7; for the value of fractions is diminished by increasing their denominators, and comes to $\frac{2}{35}$. Again; because $\frac{7}{9}$ is nine times less than seven, the quotient of any number divided by $\frac{7}{9}$ will be nine times greater than the quotient of the same number divided by 7. Therefore we multiply $\frac{2}{35}$ by 9, and obtain $\frac{18}{35}$.

If the divisor and dividend have the same denominator, it is sufficient to divide the numerators.

Ex. $\frac{12}{17}$ divided by $\frac{3}{17}$ quotes 4.

The foregoing rule may be extended to every case by reducing integers and mixed numbers to the form of improper fractions. We shall add some directions for shortening the operation when integers and mixed numbers are concerned.

1st. When the dividend is an integer, multiply it by the denominator of the divisor, and divide the product by the numerator.

Hitherto we have considered the fractions as abstract numbers, and laid down the necessary rules accordingly. We now proceed to apply these to practice. Shillings and pence may be considered as fractions of pounds; and lower denominations of any kind as fractions of higher; and any operation, where different denominations occur, may be wrought by expressing the lower ones in one form of vulgar fractions, and proceeding by the following rules. For this purpose the two following problems are necessary : —

PROBLEM V.   To reduce lower Denominations to Fractions of higher.

Rule. *Place the given number for the numerator, and the value of the higher for the denominator.*

Examples.

1. Reduce 7*d.* to the fraction of a shilling. *Ans.* $\frac{7}{12}$.
2. Reduce 7*d.* to the fraction of a pound. *Ans.* $\frac{7}{240}$.
3. Reduce 15*s.* 7*d.* to the fraction of a pound. *Ans.* $\frac{187}{240}$.

PROBLEM VI. To value Fractions of higher Denominations.

Rule. *Multiply the numerator by the value of the given
denominator, and divide the product by the denominator ;
if there be a remainder, multiply it by the value of the next
denomination, and continue the division.*

Ex. 1. Required the value of $\frac{17}{60}$ of $1l$.

$$
\begin{array}{r}
17 \\
20s. \\
\hline
60)\ 340\ (5s.\ 8d. \\
300 \\
\hline
40 \\
12d. \\
\hline
60)\ 480\ (8d. \\
480
\end{array}
$$

Ex. 2. Required the value of $\frac{8}{9}$ of 1 cwt.

$$
\begin{array}{r}
8 \\
4\ qrs. \\
\hline
9)\ 32\ (3\ qrs.\ 15\tfrac{5}{9}\ lb. \\
27 \\
\hline
5 \\
28\ lb. \\
\hline
9)\ 140\ (15\ lb. \\
9 \\
\hline
50 \\
45 \\
\hline
5 = 8\ oz.\ \tfrac{8}{9}\ oz.
\end{array}
$$

In the first example we multiply the numerator 17 by
20, the number of shillings in a pound, and divide the
product 340 by 60, the denominator of the fraction, and
obtain a quotient of 5 shillings ; then we multiply the re-

mainder, 40, by 12, the number of pence in a shilling, which produces 480, which, divided by 60, quotes 8*d*. without a remainder. In the second example we proceed in the same manner, but as there is a remainder, the quotient is completed by a fraction.

Sometimes the value of the fraction does not amount to an unit of the lowest denomination; but it may be reduced to a fraction of that or any other denomination by multiplying the numerator according to the value of the places. Thus $\frac{1}{1289}$ of a pound is equal to $\frac{12}{1289}$ of a shilling, or $\frac{240}{1289}$ of a penny, $\frac{960}{1289}$ of a farthing.

Now, however, that half-farthings are issued by authority, $\frac{1}{1289}$ of $\mathcal{L} = \frac{1920}{1289} = 1\frac{651}{1289}$ half-farthings; so that the remainder is something more than $\frac{1}{2}$ a farthing.

SECTION II.

DECIMAL FRACTIONS.

Subsect. I. — NOTATION AND REDUCTION OF DECIMAL FRACTIONS.

DECIMAL fractions are such as have 10, or some power of 10 (that is, 100, 1000, &c.), for the denominator: such are these —

$$\frac{3}{10}, \ \frac{24}{100}, \ \frac{75}{1000}, \ \frac{462}{100000}.$$

They are more simply written thus:—

·3, ·24, ·075, ·00462;

the number of figures after the point being always the same as the number of ciphers in the denominators.

In decimal fractions, as thus written, the figure next the point to the right indicates so many tenths; the next so many hundredths, and so on. Thus, in the fraction ·346, the figure 3 expresses 3-tenths, 4 denotes 4-hundredths, and 6, 6-thousandths.

The use of ciphers in decimals, as well as in integers, is, to bring the significant figures to their proper places, on which their value depends. As ciphers, when placed on the left hand of an integer, have no signification, and cannot, therefore, increase its value, but, when placed on the right hand, do increase the value ten times each; so also ciphers, when placed on the right hand of a decimal, have no signification, and cannot therefore add to its value, but when placed on the left hand, diminish the value ten times by each cipher we add thereto.

The notation and numeration of decimals will become obvious from the following examples : —

4·7   signifies four, and seven tenth parts.

·47   —   four tenth parts, and seven hundredth or 47 hundredth parts.

·047   —   four hundredth parts, and seven thousandth parts, or 47 thousandth parts.

·407   —   four tenth parts, and seven thousandth parts, or 407 thousandth parts.

4·07   —   four, and seven hundredth parts.

4·007   —   four, and seven thousandth parts.

To reduce vulgar fractions to decimal ones: *Annex a cipher to the numerator, and divide it by the denominator, annexing a cipher continually to the remainder.*

Ex. 1. $\frac{12}{75} = \cdot 16.$ Ex. 2. $\frac{5}{64} = \cdot 078125.$

$$
\begin{array}{r}
75)\,120\,(.16 \\
75 \\
\hline
450 \\
450 \\
\hline
\end{array}
\qquad
\begin{array}{r}
64)\,500\,(.078125 \\
448 \\
\hline
520 \\
512 \\
\hline
80 \\
64 \\
\hline
160 \\
128 \\
\hline
320 \\
320 \\
\hline
\end{array}
$$

In some cases, however, we find a recurrence of the same figure in the quotient, as in the division of 2 by 3, or by 6, as in the 3d and 4th examples following: in other cases the quotient will consist of figures that circulate and repeat themselves in three or two figures, as in examples 5 and 6.

Ex. 3. $\frac{2}{3} = \cdot\dot{6}66$.   Ex. 4. $\frac{5}{6} = \cdot 8\dot{3}3$.

3) 20 ($\cdot\dot{6}66$ a pure repeater.   6) 50 ($\cdot 8\dot{3}3$ a mixed re-
 18   48   peater.
 ——   ——
 20   20
 18   18
 ——   ——
 20   20
 18   18
 ——   ——
 20   20

Ex. 5. $\frac{7}{27} = \cdot 2\dot{5}\dot{9}$.   Ex. 6. $\frac{7}{22} = \cdot 3,18,18$.

27) 70 ($\cdot 25\dot{9}$ 22) 70 ($\cdot 318$ 18
 54   66
 ——   ——
 160   40
 135   22
 ——   ——
 250   180
 243   176
 ——   ——
 7   40
            22
            ——
            18

The reason of this operation will be evident, if we consider that the numerator of a vulgar fraction is understood to be divided by the denominator; and this division is actually performed when it is reduced to a decimal.

In like manner, when there is a remainder left in division, we may extend the quotient to a decimal, instead of completing it by a vulgar fraction, as in the following example: —

$$25)\ 646\ (25\tfrac{21}{25},\ \text{or } 25\cdot84\ ;\ \text{for } \tfrac{21}{25}=\cdot84$$

           50
          ———
          146
          125
          ———
    Rem. 21·0
          200
          ———
          100
          100

From the foregoing examples, we may distinguish the several kinds of decimals. Some vulgar fractions may be reduced exactly to decimals, as Ex. 1. and 2., which are called *terminate* or *finite decimals*. Others cannot be exactly reduced, because the division always leaves a remainder; but, by continuing the division, it will be perceived how the decimal may be extended. These are called *infinite decimals*. If the same figure continually returns, as in Ex. 3. and 4., they are called *repeaters*. If two or more figures return in their order, they are called *circulates*. If this regular succession go on from the beginning, they are called *pure repeaters*, or *circulates*, as Ex. 3. and 5. If otherwise, as Ex. 4. and 6., they are mixed *repeaters* or *circulates*, and the figures prefixed to them in regular succession are called the *finite part*. Repeating figures are generally distinguished by a dash, and circulates by a comma or other mark, at the beginning and end of the circle ; and the beginning of a repeater or circulate is pointed out in the division by an asterisk.

Lower denominations may be considered as fractions of higher ones, and reduced to decimals accordingly. We may proceed by the following rule, which is the same in effect as the former : —

To reduce lower denominations to decimals of higher. *Annex a cipher to the lower denomination, and divide it by the value of the higher. When there are several denominations, begin at the lowest, and reduce them in their order.*

Ex. To reduce 5 cwt. 2 qrs. 21 lb. to the decimal of a ton.

```
28) 210 (·75      4) 2·75 (·6875       20) 5·6875 (·284375
    196              24                     40
    ———             ——                     ——
    140              35                    168
    140              32                    160
    ———             ——                     ———
                     30                     87
                     28                     80
                    ——                     ——
                     20                     75
                     20                     60
                    ——                     ——
                                           150
                                           140
                                           ———
                                           100
                                           100
                                           ———
```

Here, in order to reduce 21 lb. to a decimal of 1 qr., we
annex a cipher, and divide by 28, that being the number
of lbs. in 1 qr. This gives ·75. Then we reduce 2·75 qrs.
to a decimal of 1 cwt. by dividing by 4, the number of
quarters in 1 cwt., and the quotient is ·6875. Lastly,
5·6875 cwt. is reduced to a decimal of a ton by dividing
by 20, there being 20 cwt. in a ton, and the quotient
comes to ·284375; therefore, 5 cwt. 2 qrs. 21 lb. are
decimally .284375 tons.

To find the value of a decimal fraction. *Multiply the
decimal by the value of the denomination, and cut off as
many decimal places from the product as there are in the
multiplicand. The rest are integers of the lower denomination.*

Ex. What is the value of ·425 of 1*l.*?

$$\left.\begin{array}{l} ·425 \\ \underline{20} \\ 8·500s. \\ \underline{6} \\ 3·000d. \end{array}\right\} \text{therefore, } ·425l. = 8s.\ 6d.$$

So, also, the value of ·634375 of £ = 12s. 8½d.; and
·916£ = 18s. 4d.

Subsect. II.—ARITHMETIC OF TERMINAL DECIMALS.

The value of decimal places decreases, like that of integers, ten of the lower place in either being equal to one of the next higher; and the same holds in passing from decimals to integers. Therefore, all the operations are performed in the same way with decimals, whether placed by themselves or annexed to integers, as with pure integers. The only peculiarity lies in the arrangement and pointing of the decimals.

In Addition and Subtraction, *Arrange units under units, tenth parts under tenth parts, and proceed as in integers.*

32·035	from	13·348	and 12·248
116·374	take	9·2993	less 10·6752
160·63			
12·3645	and we have	4·0487	Difference 1·5728

Sum 321·4035

In Multiplication, *Allow as many decimal places in the product as there are in both factors. If the product has not so many places, supply them by prefixing ciphers on the left hand.*

Ex. 1. 1·37   Ex. 2. 43·75   Ex. 3. ·1572
          1·8   ·48   ·12
        ─────   ─────   ─────────
         1096   35000   ·018864 prod.
          137   17500
        ─────   ───────
       2·466 prod.   21·0000 prod.

The reason of this rule may be explained by observing, that the value of the product depends on the value of the factors; and since each decimal place in either factor diminishes its value ten times, it must equally diminish the value of the product.

To multiply decimals by 10, move the decimal point one place to the right; to multiply by 100, 1000, or the

C

like, move it as many places to the right as there are ciphers in the multiplier.

In division, *Point the quotient so that there may be an equal number of decimal places in the dividend as in the divisor and quotient together.*

Therefore, if there be the same number of decimal places in the divisor and dividend, there will be none in the quotient.

If there be more in the dividend, the quotient will have as many as the dividend has more than the divisor.

If there be more in the divisor, we must annex (or suppose annexed) as many ciphers to the dividend as may complete the number in the divisor, and all the figures of the quotient are integers.

If the division have a remainder, the quotient may be extended to more decimal places; but these are not regarded in fixing the decimal point.

To find the reciprocal of any number, divide 1 with ciphers annexed by that number.

Ex. Required the reciprocal of 625.

$$625)\ 1\!\cdot\!000\ (\!\cdot\!0016$$
$$\underline{625}$$
$$3750$$
$$3750$$

The product of any number multiplied by ·0016 is the same as the quotient divided by 625.

Subsect. III.—Approximate Decimals.

It has been shown that some decimals, though extending to any length, are never complete; and others which terminate at last, sometimes consist of so many places that it would be difficult in practice to extend them fully. In these cases, we may extend the decimal to three, four, or more places, according to the nature of the articles and the degree of accuracy required, and reject the rest of it as inconsiderable. In this manner we may perform operations with ease by the common rules, and the answers we obtain are sufficiently exact for the purposes of business. Decimals thus restricted are called *approximates*.

Shillings, pence, and farthings may be easily reduced to decimals of three places by the following rule: take half the shillings for the first decimal place, and the number of farthings increased by one, if it amount to 24 or upwards; by two, if it amount to 48 or upwards; and by three, if it amount to 72 or upwards, for the next places.

The reason of this is, that 20 shillings make a pound, two shillings is the tenth part of a pound, and therefore half the number of shillings makes the first decimal place. If there were 50 farthings in a shilling, or 1000 in a pound, the units of the farthings in the remainder would be thousandth parts, and the tens would be hundredth parts, and so would give the two next decimal places; but because there are only 48 farthings in a shilling, or 960 in a pound, every farthing is a little more than the thousandth part of a pound; and since 24 farthings make 25 thousandth parts, allowance is made for that excess by adding for every 24 farthings, as directed.

If the number of farthings be 24, 48, or 72, and consequently the second and third decimal places 25, 50, and 75, they are exactly right; otherwise they are not quite complete, since there should be an allowance of $\frac{1}{24}$, not only for 24, 48, and 72 farthings, but for every other single farthing. They may be completed by the following rule: Multiply the second and third decimal places, or their excess above 25, 50, 75, by 4. If the product amount to 24 or upwards, add 1; if 48, add 2; if 72, add 3. By this operation we obtain two decimal places more; and by continuing the same operation, we may extend the decimal till it terminates in 25, 50, 75, or in a repeater.

Decimals of sterling money of three places may easily be reduced to shillings, pence, and farthings, by the following rule: Double the first decimal place, and if the second

be 5 or upwards, add 1 thereto for shillings.   Then divide
the second and third decimal places, or their excess above
50 by 4, first deducting 1, if it amount to 25 or upwards;
the quotient is pence, and the remainder farthings.

As this rule is the converse of the former one, the
reason of the one may be inferred from that of the other.
The value obtained by it, unless the decimal terminate in
25, 50, or 75, is a little more than the true value; for
there should be a deduction, not only of 1 for 25, but a
little deduction of $\frac{1}{25}$ on the remaining figures of these
places.

We proceed to give some examples of the arithmetic of
approximates, and subjoin any necessary observations.

ADDITION.					SUBTRACTION.		
cwt.	qrs.	lb.			cwt.	qrs.	lb.
3	2	14=	3·625		3	2	2=3·51785
2	3	22=	2·94642		1	1	19=1·41964
3	3	19=	3·91964		2	0	11=2·09821
4	1	25=	4·47321				
14	3	24=	14·96427				

If we value the sum of the approximates, it will fall a
little short of the sum of the articles, because the decimals
are not complete.

It is proper to add 1 to the last decimal place of the
approximate, when the following figure would have been
5 or upwards.   Thus the full decimal of 3 qrs. 22 lb. is
946·428571, and therefore ·94643 is nearer to it than
946·42.   Approximates thus regulated will give exacter
answers, sometimes above the true one and sometimes below it.

The mark + signifies that the approximate is less than
the exact decimal, or requires something to be added.
The mark — signifies that it is greater, or requires something to be subtracted.

Section III.

DUODECIMALS, or CROSS MULTIPLICATION.

Duodecimals (proceeding by twelves) is a term given
to a rule or operation of arithmetic, by which the contents
of any surface or solid are found by multiplying together
its lineal dimensions, expressed in feet, inches, and lines,
and is consequently much used by artificers in finding the
contents of their work. The rule is also called cross
multiplication, from the manner in which the operation
is usually performed, and which is as follows: suppose it
were required to find the superficial content of a plank
12 ft. $9\frac{1}{2}$ in. long, and 3 ft. 7 in. broad. Set down the
two dimensions under each other, placing feet under feet,
inches under inches, &c., and for the half inch put down
its equivalent 6 lines, as in the following example: —

ft.	in.	lin.
12	9	6
3	7	
38	4	6
7	5	$6\frac{1}{2}$
45	10	$0\frac{1}{2}$
	12	
	$120\frac{1}{2}$	

Now since the feet are conceived to be units of measure, the inches are so many 12ths of unity, and the lines
so many 12ths of a 12th, or 144th parts of unity. The
units consequently form the first column, the 12ths the
second, and the 144ths the third. Multiplying, therefore,
the first line by 3 ft. or 3 units, we get 38 ft. 4-12ths of a
foot, and 6-144ths of a foot. Next multiplying the upper
line by 7-12ths, we get first the 6 lines or 6-144ths, multiplied by 7-12ths, equal to 42-1728ths, which is equal to
$3\frac{1}{2}$-144ths. Then the 9-12ths multiplied by the 7-12ths,
give 63-144ths, which, added to the $3\frac{1}{2}$, make $66\frac{1}{2}$-144ths,
or 5-12ths and $6\frac{1}{2}$-144ths; therefore $6\frac{1}{2}$ is placed in the

thi̇rd column, and the 5-12ths carried on. Lastly, the 12 units multiplied by the 7-12ths, give 84-12ths, which, added to the 5-12ths, make 89-12ths, and this is equal to 7 units or feet and 5-12ths; consequently, 7 is placed in the first column and 5 in the second. Adding the two products together, we get 45 ft. 10-12ths of a foot, and ½-144ths of a foot. But in square or 'superficial measure the 144th part of a foot is an inch; and 10-12ths = 120-144ths; consequently, the result of the operation is 45 sq. feet and 120½ sq. inches.

The operation is itself much simpler than the description or explanation, which is found embarrassing to beginners; it would therefore, perhaps, be better to reject the rule altogether from elementary books of arithmetic; and, regarding the inches and lines as part of a foot, to perform the operation by the ordinary rules of *practice*, or, which is better, by reducing both factors to decimals, and proceeding as in simple multiplication.

Thus, the foregoing question worked decimally is $12 \cdot 7916 \times 3 \cdot 583 = 45 \cdot 8522$ feet $= 45$ feet 10 in. &c.

Section IV.

DATA FOR ARITHMETICAL CALCULATIONS.

Subsect. I.—Tables of Weights and Measures.

The following is a tabular view of the weights and measures, according to the present state (1845) of the law, throughout the British empire. In some of the colonies, however, and particularly in India, a variety of other weights and measures besides is still in use. It is only necessary to observe, that all the quantities in the same horizontal line of the same table are equal to each other.

Measures of Length.

Inches.	Links.	Feet.	Yards.	Pole or Perch.	Chains.	Fur-longs.	Mile.
7·92	1						
12	1·515	1					
36	4·545	3	1				
198	25	16·5	5·5	1			
792	100	66	22	4	1		
7920	1000	660	220	40	10	1	
63360	8000	5280	1760	320	80	8	1

Three inches make a palm, 4 inches a hand, 5 feet a pace, and 6 feet a fathom.   In cloth measure, $2\frac{1}{4}$ inches = 1 nail, 4 nails = 1 quarter, and 4 quarters = 1 yard.

Measures of Surface.

Square Inches.	Square Links.	Square Feet.	Square Yards.	Square Pole or Perch.	Square Chain.	Square Rood.	Acre.
62·726	1						
144	2·295	1					
1296	20·661	9	1				
39204	625	272·25	30·25	1			
627264	10000	4356	484	16	1		
1568160	25000	10890	1210	40	2·5	1	
6272640	100000	43560	4840	160	10	4	1

In the superficial measurement of stone, brick, or slate work, 36 square yards are termed a rood, and 100 square feet of flooring a square.   There are 1728 cubic inches in the cubic foot, and 27 cubic feet in the cubic yard; 40 cubic feet of rough, or 50 of hewn timber, make a load or ton; and 42 cubic feet make 1 ton of shipping.   A cubic yard of earth is called a load.

Imperial Liquid and Dry Measure.

Pounds of Water.	Cubic Inches.	Gills.	Pints.	Quarts.	Pottles.	Gallons.	Pecks.	Bushels.	Coombs.	Quarter.
1·25	34·659	4	1							
2·5	69·318	8	2	1						
5	138·637	16	4	2	1					
10	277·274	32	8	4	2	1				
20	554·548	64	16	8	4	2	1			
80	2218·191	256	64	32	16	8	4	1		
320	8872·763	1024	256	128	64	32	16	4	1	
640	17745·526	2048	512	256	128	64	32	8	2	1

The peck, bushel, coomb, and quarter are dry measures only.

In beer measure, the barrel consists of 36 gallons = 4 firkins, and the hogshead contains 1½ barrel or 54 gallons. The anker, tierce, hogshead, puncheon, pipe, butt, and tun, used for wine and spirits, are so vague and variable in their contents, that they are to be considered rather as the names of the casks, than as expressing any fixed or definite measures.

The Scottish ell was = 37·0598 imperial inches; and 1920 ells = 32 falls = 8 furlongs = 1 mile = 1·123024 imperial mile. The Scottish chain for land measure, like the imperial, consisted of 100 links, and though often reckoned to have been just 74 imperial feet, was more correctly = 24 ells = 74·1196 feet; and 5760 square ells = 160 falls = 10 square chains = 4 roods = 1 acre = 1·261183 imperial acre. Scottish acres will, therefore, be reduced to imperial, by multiplying them by 1·261183. Twenty-three Scottish acres made about 29 imperial, or, more nearly, 134 Scottish = 169 imperial.

In the old Scottish liquid measure, 128 gills = 32 mutchkins = 16 chopins = 8 pints = 1 gallon = 3·00651 imperial gallons.

Although several of the old Scottish dry measures were very different in different districts, most of them were similarly subdivided, having 64 lippies or forpets = 16 pecks = 4 firlots = 1 boll. In some of them, however, the bolls, especially where they were large, had very different sorts of subdivisions.

Troy Weight.

Grains.	Dwts.	oz.	lb.
24	1		
480	20	1	
5760	240	12	1

Apothecaries' Weight.

Troy Grains.	Scruples.	Drams.	oz.	lb.
20	1			
60	3	1		
480	24	8	1	
5760	288	96	12	1

In these two weights the grain, ounce, and pound are the same. The troy is used for the precious metals and for jewels, as also in trying the strength of spirituous liquors, and for comparing different weights with each other.

Four grains troy make a carat. But this term, when applied to gold, denotes its degree of fineness. Thus, the weight of any quantity or compound of that metal being supposed to be divided into 24 equal parts, if the mass be pure gold, it is said to be 24 carats fine. If it consist of 23 parts of pure gold and 1 of alloy, it is said to be 23 carats fine, and so on. Diamonds and pearls are also weighed by carats of 4 grains, but 5 diamond grains are only equal to 4 troy grains. This sort of weight is not very different all over the globe. There are 150 diamond carats in the troy ounce. Apothecaries' weight is chiefly used for medical prescriptions; but drugs are mostly bought and sold, especially in wholesale, by avoirdupois weight.

Avoirdupois or Commercial Weight.

Troy Grains.	Drams.	oz.	lb.	Stones.	Qrs.	cwt.	Ton.
437·5	16	1					
7000	256	16	1				
98000	3584	224	14	1			
196000	7168	448	28	2	1		
784000	28672	1792	112	8	4	1	
15680000	573440	35340	2240	160	80	20	1

The above pound of 7000 troy grains was formerly subdivided into 7680 avoirdupois grains, 10 of which made a scruple, 30 a dram, and 430 an ounce. The troy pound is less than the avoirdupois in the proportion of 144 to 175, or of 14 to 17 nearly; but the troy ounce is greater than the avoirdupois in the proportion of 192 to 175, or of 79 to 72 nearly. (*Encyclopædia Britannica*, 7th ed. vol. xxi. p. 850.)

Imperial measure of capacity for coals, culm, lime, fish, potatoes, fruit, and other goods, commonly sold by heaped measure : —

2 gallons make 1 peck   = 704 cubic inches, nearly.
8 gallons ...... 1 bushel = 2815¼
3 bushels ...... 1 sack   =   4⅚ cubic feet, nearly.
12 sacks .... 1 chaldron = 58⅔

The goods are to be heaped up in the form of a cone, to a height above the rim of the measure, of at least three fourths of its depth. The outside diameters of measures, used for heaped goods, are to be at least double the depth, consequently not less than the following dimensions:

Bushel............ 19½ inches.	Gallon............ 9¾ inches.
Half bushel...... 15¼	Half gallon...... 7¾
Peck.............. 12¼	

The imperial gallon contains exactly 10lbs. avoirdupois of pure or distilled water; consequently, the pint will hold 1¼ lb. and the bushel 80 lb.

Particular Weights.

8 pounds make 1 stone, used for meat.

14 pounds ...... 1 stone = 0 cwt. 0 qrs. 14 lb. ⎫
2 stone .... 1 tod = 0   1   0 |
6½ tod .... 1 wey = 1   2   14   ⎬ used in the wool trade.
2 weys .... 1 sack = 3   1   0 |
12 sacks .... 1 last = 39   0   0   ⎭

Miscellaneous.

12 dozen make a gross.
A weigh is 256 lbs.
12 barrels make a last.
A quire of paper is 24 sheets.
A ream of paper is 20 quires.
A bundle of paper is 2 reams.
A bale of paper is 10 reams.
A roll of parchment, or vellum, is five dozen, or 60 skins.
A dicker of hides is 10 skins.
A last of hides is 20 dickers.
A dicker of gloves is 10 dozen pairs.
A firkin of butter is 56 lbs.
A firkin of soap is 64 lbs.
A tierce of rice is about 5 cwt.
A hogshead of tobacco is from 9 to 10 cwt.
A barrel of gunpowder is 1 cwt.

A pack of wool is 240 lbs.
20 stones of flour make a sack.
A load of timber, unhewed, is 40 feet.
A load of bricks, 500 in number.
A load of tiles, 1000 in number.
A load of hay, in London, is nearly 18 cwt.
A load of straw, 36 trusses, of 36 lbs. each.
A chaldron of coals, in London, is 36 bushels. *Coals are sold by the ton.*
A chaldron of coals, in Newcastle, is 53 cwt.
A cart of coals, in Scotland, is 12 cwt
A deal of coals, in Scotland, is 23 cwt.
A grain of gold is worth about 2d.
An ounce of fine silver is worth from 5s. to 5s. 6d.

Measures of Time.

Time is a measure of duration, ascertained by the motions of the heavenly bodies, and is either apparent, mean, or sidereal.

The Julian year is equal to 365 days 6 hours.
Solar year...................... 365 days, 5 hours, 48 min. 48 sec.
Civil year...................... 365 days, or 12 calendar months.
Lunar month .... 4 weeks, or 28 days.
Civil, or natural day ...... 24 hours.
An hour .... 60 minutes.
A minute .... 60 seconds.

Subsect. II.—Rule for calculating Interest at 5 per cent.

Multiply the pounds by the days, and divide the product by 365. The quotient gives the interest at 5 per cent. in shillings.

Subsect. III.—Table to calculate Wages and other Payments.

Y.	Per Month.			Per Week.			Per Day.		Y.	Per Month.			Per Week.			Per Day.	
£	£	s.	d.	£	s.	d.	s.	d.	£	£	s.	d.	£	s.	d.	s.	d.
1	0	1	8	0	0	4¾	0	0¾	15	1	5	0	0	5	9	0	10
2	0	3	4	0	0	9¼	0	1¼	16	1	6	8	0	6	1¾	0	10½
3	0	5	0	0	1	1¾	0	2	17	1	8	4	0	6	6¼	0	11¼
4	0	6	8	0	1	6½	0	2¾	18	1	10	0	0	6	10¾	0	11¾
5	0	8	4	0	1	11	0	3¼	19	1	11	8	0	7	3½	1	0½
6	0	10	0	0	2	3½	0	4	20	1	13	4	0	7	8	1	1¼
7	0	11	8	0	2	8¼	0	4½	30	2	10	0	0	11	6	1	7¾
8	0	13	4	0	3	0¾	0	5¼	40	3	6	8	0	15	4	2	2¼
9	0	15	0	0	3	5½	0	6	50	4	3	4	0	19	2	2	9
10	0	16	8	0	3	10	0	6½	60	5	0	0	1	3	0¼	3	3¼
11	0	18	4	0	4	2¾	0	7¼	70	5	16	8	1	6	10¼	3	10
12	1	0	0	0	4	7¼	0	8	80	6	13	4	1	10	8¼	4	4½
13	1	1	8	0	4	11¾	0	8½	90	7	10	0	1	14	6¼	4	11½
14	1	3	4	0	5	4¼	0	9¼	100	8	6	8	1	18	4½	5	5¾

If the wages be guineas instead of pounds, for each guinea add one penny to each month, or one farthing to each week.

Subsect. IV.—Quarterly Terms.

First term, or *Lady-day*, falls on the 25th March.
Second term, or *Midsummer* . . . 24th June.
Third term, or *Michaelmas* . . . 29th September
Fourth term, or *Christmas* . . . 25th December.

Note.—In Scotland, the first quarter term is on *Candlemas*, or the 2d of February; the second term, *Whitsunday*, is on the 15th of May; the third term, or *Lammas*, on the 1st of August; and the fourth term, *Martinmas*, on the 11th of November.

Section V.

PROPORTION AND INTEREST.

Subsect. I. — Of Proportion, or the Rule of Three.

Proportion is that rule by which from a comparison of circumstances, arising from certain conditions or stipulations, certain conclusions are drawn, and consequences deduced and ascertained.

In simple proportion, three numbers are always given to find a fourth ; of which the first two are always conditional, and the third implies a demand, and in consequence moves the question. In all direct processes the answer, or fourth proportional, bears the same ratio to the third as the second bears to the first ; wherefore, the greater the second term is in respect to the first, the greater will the fourth term be in respect to the third ; and the less the second term is in respect to the first, the less will the fourth term be in respect to the third. Hence in all direct proportions, the product of the extremes will always be equal to the product of the means. On the other hand, if the terms are in reciprocal proportion, the fourth proportional must always bear the same ratio to the second as the third does to the first ; consequently, the greater the third term is in respect of the first, the less must the fourth be in respect to the second ; and the less the third is, compared with the first, the greater will the fourth be, compared with the second. Hence again, in reciprocal proportion, the product of the first and second terms will always be equal to the product of the third and fourth.

To state the question.

Rule I. Write down that number or term which is of the kind, whether money, weight, measure, time, &c. with the answer, for the middle term.

Rule II. On the right of the middle term already written down, place the term upon which the demand lies.

Rule III. On the left of the middle term, place that term of the two conditional ones which is of the same kind with the term on the right. Then will the terms be placed in a proportional order.

Or, write down that term on which the demand lies for the middle term, and the term homogeneal with it, on the left, and the term homogeneal with the answer on the right.

To find a fourth proportional.

Rule. If, upon comparing the first and third, more be found to require more, or less to require less, then will the terms be in direct proportion, and the product of the two last, divided by the first, will quote the answer ;٭but if less require more, or more require less, then will the product of the two first, divided by the last, quote the answer.

Note. — Similar terms, that is, the first and third, or second and fourth, must always be of the same denomination, and the preparation may be made as in Reduction, Vulgar or Decimal Fractions.

Ex. 1. Bought 175½ yards of cloth, and paid 105*l.* 12*s.* 0*d.* what did it cost me per piece of 25 yards?

```
 yds.              £                     yds.
175½     :      105·6        ::           25
 351             50                        2
            351) 5280·0 (15 . 0 . 10 50/117    50
                 351
                 1770
                 1755
                   15
                   20
                  300
                   12
                 3600
                  351
                   90
```

Illustration. Because the answer is to be money, the price is put for the middle term, on the right of which stands the term which implies a demand, 25 yards, and on the left 175½ yards, being the term which is of the same kind with 25. Now, as 25 yards, when compared with 175½, reckon less, and of consequence must bring less, the two last terms are multiplied together, and their product divided by the first. Previous to any multiplication or division, because in the first term there is a fraction, the first and third terms are reduced into parts expressed by the denominator; and since there are shillings in the second, as well as pounds, the shillings are reduced to a decimal.

The same answer may be effected decimally, thus:

$$175 \cdot 5 \ : \ 105 \cdot 6 \ :: \ 25 : 15 \ . \ 0 \ . \ 10\tfrac{1}{3} \text{ nearly}$$

$$
\begin{array}{r}
25 \\
\hline
5280 \\
2112 \\
\hline
\end{array}
$$

$$175 \cdot 5) \ 26400 \ (15 \cdot 042$$

$$
\begin{array}{r}
1755 \\
\hline
8850 \\
8775 \\
\hline
7500 \\
7020 \\
\hline
4800 \\
3510 \\
\hline
1290 \\
\hline
\end{array}
$$

In questions of this character, the product of the extremes are equal to the product of the mean terms; but in the present question we must consider that the 25 yards constitute but 7·02 parts of the whole quantity; therefore the proof is effected by multiplying the 4th term, 15*l.* 0*s.* 10½*d.*, by 7·02, and the product will be 105*l.* 12*s.* nearly.

Or by reduction, thus :

$175\frac{1}{2}$　:　105·12　::　25
　　　　　　　　20
　—————
351　　: 2112　　::　50
　　　　　　50
　　—————
351) 105600 (300s. $10\frac{30}{117}d.$ = 15l. 0s. $10\frac{2}{3}d.$
　　1053
　　—————
　　　　300
　　　　 12
　　　—————
　　　3600
　　　 351
　　　—————
　　　　 90

Or by vulgar fractions, thus :

$\frac{351}{2} : \frac{2112}{10} :: \frac{25}{1} : \frac{105600}{7020} = 15\frac{30}{702} = 15l.\ 0s.\ 10\frac{1}{3}d.$

Or thus :　　　　　105　　12
　　　　　　　　　　　　10
　　　　　　————
　　　　　　1056
　　　　　　　 5
　　　　　　————
351) 5280
　　　15　 0　$10\frac{1}{3}$ as before.

This question by the second arrangement would stand

yds.　yds.　£　　　£
175·5　: 25 :: 105·6　: 15·042.

But the work, as well as the answer, will be the same as before.

The work of all questions in a direct proportion may be readily proved by multiplying extremes and means, the products of which, when the work is right, will be respectively equal; but in reciprocal proportion, the product of the first and second terms will always be equal to that of the third and fourth.

2. Sent my neighbour upon an emergency, 217*l*. 10*s*.
for 112 days; how long may I retain 870*l*. of his money
to be indemnified ?

$$£ \qquad \text{days.} \qquad £$$

217 10 : 112 :: 870 state.

217·5 : 112 :: 870 preparation.

$$\overline{43\cdot5} \qquad\qquad \overline{174} \text{ abridged by 5.}$$

112

$$\overline{870}$$

4785

174) 48720 (28

348

$$\overline{1392}$$

1392   Ans. 28 days.

Illustration. By comparing the first and third terms
together, we find that more requires less, because it would
not be fair to keep 870*l*. the same time that its owner had·
217*l*. 10*s*., wherefore the product of the two first, divided
by the last, quotes the answer.

3. The height of my staff from the ground is 5 ft. 9 in.,
and it casts a shadow of 6 ft. 3 in.; what should be the
height of a steeple which casts a shadow of 217 ft. 6 in.?

6·25 : 5·75 :: 217·5

$$\overline{1\cdot25} \qquad \overline{1\cdot15} \qquad \overline{43\cdot5} \text{ abridged by 5 continually.}$$

·25 ·23 ·23

$$\overline{\cdot05} \qquad\qquad \overline{1305}$$

870

$$\cdot05) \overline{10\cdot005}$$

feet 200·1 answer.

(*Gordon's Arithmetic,* p. 110.)

Subsect. II. — Of Compound Proportion.

When questions in proportion require two or more operations before the answer can be obtained, as some of the foregoing rules, the process will be less perplexed if the terms are arranged in a successive order, and so reduced to one simple operation; for which observe the following rules.

I. Of the given terms, three are conditional, and two imply a demand. Of the conditional terms, let that which is the principal cause of gain or loss, action or passion, increase or decrease, appear as the first term; let that which includes in it time or distance appear in the second place; and the remaining conditional term in the third place. The other two terms, which imply a demand, take the order of their arrangement from the other three.

II. Consider to which of the three conditional terms, as antecedents, the answer, or sixth proportional, is to be a consequent; or, by comparing the antecedents and consequents on the five terms, find which of the antecedents wants a consequent.

III. Then, if the term sought be of the same kind with the third, the continued product of the three last terms, divided by the product of the two first, will quote the answer; but if the term sought be of the same kind with the first or second, the continued product of the first, second, and fifth terms, divided by the product of the third and fourth terms, will quote the answer.

Examples.

1. If 12 roods of grass be cut down by 2 men in 6 days; how many roods will be cut down by 8 men in 24?

Men.   Days.   Roods.   Men.

$$2 \;:\; 6 \;::\; 12 \;:\; 8 \;:\; 24 \text{ days.}$$

$$\frac{12 \times 8 \times 24}{12} = 192 \text{ roods.}$$

Or, by abridging the terms:

$$8 \times 24 = 192.$$

D

Illustration. As men are .the cause of action, 2 stands in the first place; as days imply time, 6 possesses the second place; and roods being the action, 12 possesses the third place; in which order also the other two terms fall to be placed, the men first, and then the days: then, because a term of the same kind with the third is required, the continued product of the three last is divided by the product of the first two.

2. Suppose 100*l.* would defray the expense of a certain work for 22 weeks 6 days, when 5 men were employed in it: in what time would 12 men employed in the same work draw 150*l.* ?

$$\text{M.} \quad \text{D.} \quad \pounds \quad \text{M.} \quad \pounds$$
$$5 : 160 :: 100 : 12 : 150$$
$$\text{Abridged, } 5 : 16 :: 1 : 12 : 15.$$
$$\frac{5 \times 16 \times 15}{12} = 100 \text{ days.}$$

Because a term of the same kind with the second is required, the continued product of the first, second, and last, divided by the product of the third and fourth, quotes the answer.

3. What principal sum will gain 20*l.* in 3 months, at 5 per cent. per annum?

$$\pounds \quad \text{M.} \quad \pounds \quad \text{M.} \quad \pounds$$
$$100 : 12 :: 5 : 3 : 20$$
$$20 : 4 :: 1 : 1 : 20.$$
$$20 \times 4 \times 20 = 1600.$$

Here the required term is of the same kind with the first.

(*Gordon's Arithmetic,* p. 120.)

Subsect III. — SIMPLE INTEREST.

By simple interest it is to be understood that the sum paid for the use of the principal becomes at no time a part of that principal.

To find the interest for any given sum, at any rate per cent., and for any time.

I. If the time be any number of complete years, any aliquot part of a year, or both, compute by either of the following methods: —

As 100 is to the product of the rate and time, so is the principal to the interest: or, as 1*l.* to the amount of 1*l.*, at the rate, and for the time given, so is, &c. Or, multiply the rate by the time, and compute by practice for the product, as in commission.

1. What is the interest of 578*l.* 19*s.* 0*d.* for 4½ years, at 5 per cent. ?

$$\begin{array}{r} 578{\cdot}95 \\ 4\tfrac{1}{2} \\ \hline 2315{\cdot}80 \\ 289{\cdot}475 \\ \hline 2605{\cdot}275 \\ 5 \\ \hline 100)\,13026{\cdot}375 \\ \hline 130{\cdot}26375 \end{array}$$

$$\begin{array}{r} 578{\cdot}95 \\ 4{\cdot}5 \\ \hline 289475 \\ 231580 \\ \hline 2605{\cdot}275 \\ {\cdot}05 \\ \hline 130{\cdot}26375 = 130l.\ 5s.\ 3\tfrac{1}{4}d. \end{array}$$

$$\begin{array}{r} 5)\,578{\cdot}95 \\ \hline 4\tfrac{1}{2} \qquad\quad 8)\,115{\cdot}79 \text{ for 20 per cent.} \\ 5 \qquad\qquad 14{\cdot}47375 \text{ for } 2\tfrac{1}{2}. \\ \hline 22\tfrac{1}{2} \qquad\qquad 130{\cdot}26375 \text{ for } 22\tfrac{1}{2}. \end{array}$$

II. If the time be any number of days less than a year, multiply the given sum by the number of days, and divide the product by 7300, and the quotient will be the interest at 5 per cent., which may be increased or diminished to any other rate, by multiplying ⅓ of the interest found into the given rate. Or, by multiplying the quotient into the given rate, and dividing by 5.

Note. — 7300 becomes a constant divisor, because the interest of 100*l.* for 73 days, or of 73*l.* for 100 days, at 5 per cent. is just one pound.

1. What is the interest of 378l. 14s. 0d., from the 9th of November to the 16th of March, at 5 per cent. ?

Nov. 21	378·7	378·7
Dec. 31	127	·3479 = $\frac{127}{365}$
Jan. 31	26509	1·3174973
Feb. 28	45444	5
Mar. 16	73) 480·949	6·5874865 as before.
127	6·588 = 6l. 11s. 9d.	

Suppose the rate had been at 4½ per cent.

5) 6·588   10) 6·588
1·3176   ·6588 to be deducted.
      4½   5·9292 as before.
5·2704
·6588
5·9292

(*Gordon's Arithmetic*, p. 196.)

Subsect IV. — COMPOUND INTEREST.

IF a man lend 100l. for one year, and exact the payment of principal and interest when due, he will receive 105l. : if he lends out this money to the first holder, or any other, for another year precisely, he will receive back 110l. 5s., and in three years it would be 115l. 15s. 3d., &c. Hence arises compound interest, which, though it is prohibited by the laws, every banking company exact in effect, as they take particular care that no cash account shall remain unsettled beyond a year. This section also admits of four varieties, which we shall not illustrate with many examples, as all questions which occur in this rule are briefly answered by the Tables to be found at the end of the work.

Multiply the amount of 1l. for a year so often into itself as are years proposed, abating one, and the last product, multiplied by the principal, gives the amount; from which deduct the principal, for the interest.

If there are also days beyond complete years, add to the amount, or principal and compound interest formerly found, the simple interest of that amount for them.

1. What is the interest and amount of 700*l*. for $3\frac{1}{2}$ years, at 5 per cent. per annum?

Or practically,

```
      1·05                          20) 700
      1·05                               35
    ───────                        ───────────
    1·1025                         20) 735 1st year.
      1·05                              36·75
   ─────────                       ─────────────
   1·157625                        20)771·75 2d year.
      700                              38·5875
─────────────                      ─────────────
40)810·3375 amount at 3 years.     40)810·3375
   20·2584375                          20·2584375
─────────────                      ─────────────
830·5959375 amount at 3½ years.    830·5959375
```

(*Gordon's Arithmetic*, p. 205.)

Section VI.

ANNUITIES AND FREEHOLDS.

ANNUITIES are periodical payments to persons for a term of years, for life, or for ever. Thus, the dividends on stock in the public funds are annuities for unlimited terms, except otherwise expressed, as the Government "Long Annuities," which terminate in 1860. The rents of freehold estates are also annuities. The reversions of leases, after the expiration of under-leases, are deferred annuities. So, also are the reversions to freeholds, after the expiration of terminable leases. Hence we have *immediate annuities, perpetual annuities, deferred annuities,* and *deferred perpetuities.*

Annuities certain are of two kinds; first, those which are forborne, or in arrears; secondly, those which hereafter become payable at some *future* time. Respecting those in

arrear or said to be forborne, we regard only the *amount*.
When they hereafter become payable, we contemplate
their *present worth*. The former exceeds the sum of
the several payments by the *interest* which has accumu-
lated: the latter falls short of that sum by the *discount* to
be deducted. Both, however, depend, not only on the
intervals at which the payments successively become due,
but also on the periods at which the interest is convertible
into principal. Calculations of this sort are easily managed
by means of the Tables we give at the end of the work,
the construction of which we shall here illustrate by examples.

An annuity is in arrear 6 years; *i. e.* the first payment
became due 5 years ago; the second 4 years ago; the third
3; the fourth 2; the fifth 1; and the sixth is just due.

The interest is 4 per cent.; and the several payments,
as shown by the Table, are, upon 1*l.*, as follow : —

Payment just due, owes no interest, and is £1·00000	
The payment due 1 year back - -	1·04000
Their sum - - - -	2·04000
The payment due 2 years ago - -	1·08160
The sum of the three payments is -	3·12160
The payment due 3 years ago - -	1·12486
The sum of the four payments is -	4·24646
The payment due 4 years ago is - -	1·16985
The sum of these five payments is -	5·41631
The payment due 5 years ago is -	1·21665
The sum of all the six payments -	6·63296

Therefore this total is called the *years' purchase*, and is
6*l.* 12*s.* 8*d.* Hence the amount of 1*l.* per annum for 2
years is 2·04*l.*; for 3 years it is 3·1216*l.*; for 4 years it is
4·2464*l.*; for 5 years it is 5·4163*l.*; and for 6 years it is
6·632*l.*, or 6*l.* 12*s.* 8*d.*

When the annuity is made payable at the beginning of the year, each payment, and consequently their aggregate, is increased by

ANNUITIES AND FREEHOLDS.

one year's interest; so that the amount of

the last payment would be	-	- £1·04000
Of that preceding it	-	- 1·08160
Of the one preceding this	-	- 1·12486
Of the fourth preceding -	-	- 1·16985
Of the fifth preceding -	-	- 1·21665
Making a total of	-	- £5·63296

In which case we see that the annuity payable at the beginning of the year is less by unity than an annuity for 6 years payable at the end of the year.

Ex. Let the annuity be 20*l.*, forborne for 20 years at 4 per cent. compound interest. Then by the Table the amount of 1*l.* for 20 years being, at 4 per cent., 29·77808*l.*, we multiply this by 20*l.*, the amount of the annuity, and the product, 595*l.* 11*s.* 3*d.*, is the amount of the annuity.

Present worth of Annuities. — Here we consider, in addition to the above, the period when the annuity is to commence *, and the term for which it is to continue; as explained in the beginning of this article.

Thus, for the present value of an annuity of one pound due one year hence at 4 per cent., we have

The present value of 1*l.* or 1 − ·038462		
discount = -	- -	- £0·961538
To which adding that of 1*l.* due 2 years hence		0·924556
Their sum is the present value of 1*l.* per		
annum for 2 years -	-	- 1·886094
To which adding the value of 1*l.* due 3		
years hence -	- -	- 0·888996
And we have the present value of 1*l.* per		
annum for 3 years -	-	- 2·775090
Adding to this sum the value of 1*l.* due 4		
years hence -	- -	- 0·854804
The total shows the value of 1*l.* per annum		
for 4 years -	- -	- 3·629894

* An annuity payable yearly is said to commence, or be entered upon, one year before the first payment becomes due; and an annuity payable by half-yearly instalments is said to commence half a year before the first instalment becomes due; and so on.

D 4

Hence the construction of a table of present values of 1*l.*
annuity due for any number of years from 1 year up to
100 years.

If the annuity be payable at the beginning of the year, the present value of the first payment would evidently be -	£1·00000
That of the second, discounted for 1 year at 4 per cent., would be - -	0·961538
That of the third, discounted for 2 years at 4 per cent., would be - -	0·924556
That of the fourth, discounted for 3 years, at 4 per cent., would be - -	0·888996
So that the total would be -	£3·775090

Showing the present worth of 1*l.* annuity payable at the
beginning of the year for 4 years, which is simply unity
added to the present value of a like annuity payable at the
end of the year for 3 years. The same principle applies
to any other period, and to any other rate of interest : *i.e.*
adding unity to the tabular number opposite 1 year *less*
than the given term.

Ex. 1. Thus, if the improved rent of a farm held under
a lease for 21 years be 100*l.*, reckoning interest at 5 per
cent., the present value of the lease is 12·82153 × 100 =
1282*l.* 3*s.* 1*d.*

Ex. 2. And the present value of a perpetual annuity of
30*l.*, reckoning interest at 4 per cent., is 25·00000 × 30 =
750*l.*

Ex. 3. In like manner an estate in *fee-simple*, yielding a
net annual rent of 500*l.*, is, at 4 per cent., worth 25 × 500,
or 12,500*l.*

Ex. 4. Two persons, A. and B., divide an annuity of
100*l.* for the next 30 years between them, so that A. and
his heirs enjoy it for the next 10 years, and B. and his
heirs for the remaining 20 years. Required the present
value of B.'s share or *deferred* annuity, reckoning interest
at 3 per cent. ?

Under 3 per cent. in the Table, and opposite 30 years, we find the value of the
shares of A. and B. to be - - 19·60044
From which deducting A.'s share, found
in the same Table, opposite 10 years - 8·53020

And there remains B.'s share - - 11·07024
Or the years' purchase, which multiplied by 100

Produces the present value required - £1107·024

Ex. 5. What premium should be paid for adding 21
years to a lease of 50*l.* a year, of which 12 years are un-
expired, reckoning interest at 4 per cent. ?

Since 21 + 12 = 33, opposite this number in
the Table is - - - 18·14764
And opposite 12, same Table, and 4 per cent. is 9·38507

So that the years' purchase, which is - 8·76257
Multiplied by 50, the annual rent - 50

Gives 438*l.* 2*s.* 7*d.* as the premium or fine £438·12850

Ex. 6. The present value of the reversion to a freehold
estate of 300*l.* per annum, to be entered upon 15 years
hence, reckoning interest at 5 per cent. is thus determined : —

From the perpetuity $= \frac{100}{5} =$ - - 20·0000
Deduct the value of 1*l.* per annum for 15
years - - - - - 10·37965
And there remains for the reversion in
years' purchase - - - 9·62035
Which we multiply by the yearly rental, or 300
And get the present value - - £2886·105
or 2886*l.* 2*s.* 1*d.*

Proof. 2886·105*l.* laid out and improved for 15 years
would amount to (2·07892 × 2886·105 =) 5999·9814*l.*, or
say 6000*l.*, the interest on which would, at the same rate,
5 per cent., produce an annual income of 300*l.* thenceforth
without any diminution of the principal.

All the propositions and rules for freehold estates may be expressed in words, thus: —

Case I. To find the value of a freehold estate —
Multiply the yearly rent by 100, divide the product by the rate of interest, and the quotient is the answer; or, as the rate of $1l.$ is to £, so is the rent to the value sought.

Thus, rent $210l.$ a year, interest $6\frac{1}{2}$ per cent., value of the farm $= \dfrac{210 \times 100}{6 \cdot 5} = \dfrac{21000}{6 \cdot 5} = 3230l.\ 15s.\ 5d.$

Case II. To find the annual rent which a freehold estate ought to produce, so as to allow the purchaser a given rate of interest for his money:
Multiply the purchase-money by the proposed rate per cent., and divide by 100; the quotient will be the yearly rent.

Thus, the estate cost $10,000l.$, and yielded 5 per cent., therefore the annual rent would be equal to $\dfrac{10000 \times 5}{100} = 500l.$; or £1 : .05 :: 10000 : 500l.

Case III. To find what rate of interest a purchaser draws from a freehold estate:
Multiply the annual rent by 100; divide the product by the sum paid, and the quotient will be the rate per cent.

Thus, the rent is $500l.$ a year; therefore the rate of interest is equal to $\dfrac{500 \times 100}{10000} = 5$ per cent.

Case IV. To find how many years' purchase ought to be given for a freehold estate, to allow the purchaser a certain rate per cent:
Divide 100 by the proposed rate of interest, and the quotient is the number of years' purchase.

Thus, $100 \div 5 = 20$ years' purchase. Hence the construction of the following table: —

Per cent.			Per cent.		
3	is equal to	33⅓ yrs. purchase.	6	is equal to	16⅔ yrs. purchase.
3¼	—	30	6⅔	—	15
3½	—	28⁴⁄₇	7	—	14²⁄₇
4	—	25	7½	—	13⅓
4½	—	22²⁄₉	8	—	12½
5	—	20	8½	—	12
5½	—	18²⁄₁₁	9	—	11⅑

Case V. To find what rate of interest per cent. should be received for money by purchasing at so many years of annual rent:

Divide 100 by the number of years' purchase, and the quotient is the rate of interest.

Thus, an estate at 20 years' purchase yields 5 per cent., one at 25 years' purchase 4 per cent., one at 33⅓ years' purchase 3 per cent. per annum.

Of Bartering Interests in Property.

In all cases where different kinds of property are bartered, the calculations must be conducted upon the self-evident principle of equalising the present values of the properties exchanged by the following

Rule. Divide the present value per pound of the property which has its amount given by the present value of 1*l.* of that for which it is proposed to be exchanged, and the quotient multiplied by the given amount will be the answer.

Ex. To exchange a deferred perpetuity of 600*l.* to be entered upon 12 years hence for an immediate perpetuity, if the interest of money be 3½ per cent., we must give A., entitled to the reversion, 397*l.* 1*s.* 5*d.* perpetuity, and to B. 202*l.* 18*s.* 7*d.* as a deferred perpetuity.

For the present value of A.'s deferred annuity is equal to $600 \times (28 \cdot 57142 - 9 \cdot 66333) = 600 \times 18 \cdot 90809 = 11,344l.$ 17*s.* 2*d.* And since the present value of 1*l.* perpetuity is 28·57142, to determine what perpetual annuity A. can purchase with 11·344857*l.*, we say,

As £28·57142 : 1 :: £11344·857 : £397 1*s.* 5*d.* for the equivalent perpetuity as stated above.

	£	s.	d.
At 4 per cent. B. ought to give A. an immediate annuity of - -	374	15	2

And A. ought to give B. a deferred per-   £   s.   d.
petuity of   -   -   - 225   4   10
Were the interest 5 per cent., B. ought to
  give A. an immediate annuity of   - 334   2   6
And A. ought to give B. a deferred perpetuity of   -   -   - 265   17   6

Ex. 2. To determine what annuity for the next 20 years is an equivalent exchange for 5000*l.* to be received at the end of 8 years; computing by 4 per cent. By the Table we find the present value of 1*l.* due 8 years hence to be ·73069, and the present value of 1*l.* per annum for 20 years is 13·59032*l.* Therefore, dividing the first tabular number by the last, we have $\dfrac{\cdot 73069}{13\cdot 59032}=\cdot 0537655$, which, multiplied by 5000, produces 268·8275*l.* =268*l.* 16*s.* 6½*d.* for the annuity required.

Section VII.

INVOLUTION AND EVOLUTION.

If a quantity be continually multiplied by itself, it is said to be involved or raised; and the power to which it is raised is expressed by the number of times the quantity has been employed in the multiplication.

Thus, the product of any number multiplied by itself is called a *square;* as $2 \times 2 = 4 =$ the square of 2.

For this reason, the number 2, considered in relation to the product 4, is called the *square root* of 4.

Hence we have these Roots and Powers of Numbers.

Roots -	1	2	3	4	5	6	7	8	9	10
Square	1	4	9	16	25	36	49	64	81	100
Cube -	1	8	27	64	125	216	343	512	729	1000
4th -	1	16	81	256	625	1296	2401	4096	6561	10000
5th -	1	32	243	1024	3125	7776	16807	32768	59049	100000

The squares of fractions are found in the same manner, by involving the fraction into itself: thus,

Roots and Powers of Fractional Numbers.

Root	-	$\frac{1}{4}$	$\frac{1}{2}$	$\frac{3}{4}$	$\frac{1}{3}$	$\frac{2}{3}$	$\frac{1}{5}$	$\frac{2}{5}$	$\frac{3}{5}$	$\frac{4}{5}$
Square	-	$\frac{1}{16}$	$\frac{1}{4}$	$\frac{9}{16}$	$\frac{1}{9}$	$\frac{4}{9}$	$\frac{1}{25}$	$\frac{4}{25}$	$\frac{9}{25}$	$\frac{16}{25}$
Cube		$\frac{1}{64}$	$\frac{1}{8}$	$\frac{27}{64}$	$\frac{1}{27}$	$\frac{8}{27}$	$\frac{1}{125}$	$\frac{8}{125}$	$\frac{27}{125}$	$\frac{64}{125}$

OF THE SQUARE ROOT.

Evolution, or the extracting of roots, is, as we have stated above, the converse of involution. Hence if the square root of a number be sought, since we know that $10 \times 10 = 100$, the square root of 100 is therefore 10; — the $\sqrt{1000} = 100$; the $\sqrt{1000000} = 1000$; it therefore appears that the square root of any number less than 100 must consist of one figure only; — the square root of any number between 100 and 10000 must consist of 2 places of figures : of any number from 10000 to 1000000 of 3 places of figures, &c. If, then, a point be placed over every second figure in any number beginning with the units, the number of points will show the number of figures or places in the square root. Thus, the square root of $6\dot{1}15\dot{2}4$ consists of 3 figures. The operation proceeds as follows : —

$$
\begin{array}{r|l}
 & 6\dot{1}15\dot{2}4 \\
 & 49 \\
\hline
148 & 1215 \\
 & 1184 \\
\hline
1562 & 3124 \\
 & 3124 \\
\hline
\end{array}
$$

Here $7 \times 7 = 49$ we place under 61, and bring down the difference 12, to which we annex the next period 1$\dot{5}$. We find $7 + 7 = 14$, the new divisor contained 8 times in the subtrahend 121 ; and this 8, the second figure of the quotient, is placed after 14, the constant divisor, which is now 148 : then, $148 \times 8 = 1184$, which taken from 1215 leaves 31, to which we bring down the period of figures 2$\dot{4}$ first pointed off. To 148 we add 8, and get 156, which is the

new divisor, and is contained twice in 312; therefore 2
takes the third rank in the quotient, and in the constant
divisor it takes the fourth rank, which now becomes 1562;
and this multiplied by 2 gives 3124. The quotient 782
is therefore the square root of 611524. For conversely
782 × 782 = 611524.

In extracting the square root of a decimal, the first dots
must be placed over the hundredths, the second over the
tens of thousands, &c., of the *decimals*, by adding ciphers
if necessary. Thus, 64·8530 has for its square root 8·053;
and the square root of 2·7120 = 1·646, &c.

Hence, we have these rules for finding the square root.

Rule 1. Divide the given number into periods of two
places each, by placing a dot over every second figure,
beginning at the unit's place: the number of periods shows
the number of places in the root.

2. Find by trials the nearest root of the first period,
which set in the quotient, subtract its square from that
period, bring down the next period to the remainder for a
new dividend.

3. Double the quotient for a new divisor, see how often
it can be had in the dividend without the unit's place, and
annex the quotient figure both to the new divisor and to
the former quotient.

4. Multiply the divisor thus increased by the last found
figure in the quotient, and subtract the product from the
dividend. Bring down the next period to this remainder,
and proceed as before.

Applications of the Square Root.

First. To find a mean proportional between two numbers. *Rule.* Multiply the two given numbers together,
and extract the square root of the product, which will be
the mean proportional sought. Thus, the mean proportional between 3 and 12 is 6.

Ex. In a pair of scales a cheese weighed 64 lbs. in one

scale and only 49 lbs. in the other; therefore its true weight
was $\sqrt{49 \times 64} = \sqrt{3136} = 56$ lbs.

Secondly. To find the side of a square of equal content
with any given superficies. *Rule.* Extract the square root
of the number expressing the superficies of the given figure.

Ex. Thus, a garden in the form of a square which shall
contain exactly an acre of ground has its side equal to
$\sqrt{43560}$ feet $= 208 \cdot 71$ feet.

Thirdly. To increase or diminish a circle in any given
proportion. *Rule.* Square the diameter, enlarge or diminish it in the proportion required, and extract the square
root thereof.

Ex. The diameter of the imperial bushel is $19\frac{1}{2}$ inches,
and its depth $8\frac{1}{2}$ inches; therefore a bushel measure whose
depth is only $7\frac{1}{2}$ inches would require to be $20 \cdot 18$ inches
diameter.

Fourthly. To find any side of a right-angled triangle.
1. For the hypothenuse or slant side, extract the square
root of the sum of the base and perpendicular. 2. For the
perpendicular, extract the square root of the difference of
the squares of the hypothenuse and base. 3. For the
base, extract the difference of the hypothenuse and perpendicular.

OF THE CUBE ROOT.

We showed by the foregoing table that the third power
of 10 is 1000, therefore 10 is the cube root of 1000. And
the cube root of 64 is 4. Hence the following rules for
finding the cube root of numbers : —

Rule 1. Divide the given numbers, as 262144, into
periods of three figures, beginning at the unit's place, and
pointing to the left in integers, and to the right in decimals.

2. Find the nearest root in the first period, which set
in the quotient, subtract its cube from the period, and
bring down the next period to the remainder for a new
dividend.

3. Find a divisor by multiplying the square of the part of the root found by 300: divide the dividend by it, and place the quotient figure for the second figure in the root.

4. Multiply the part of the root formerly found by the last figure placed in the root, and this product by 30; which place under the divisor, under which write the square of the figure last placed in the root.

5. Multiply the sum of these three by the figure last placed in the root, and subtract the product from the dividend: to the remainder annex the next period for a new dividend, with which proceed as before; or, thus abbreviated,

$$
\begin{array}{r|l}
 & 405\overset{\cdot}{2}2\overset{\cdot}{4}\,(74 \\
7 \times 7 \times 7 = & 343 \\
7 \times 7 \times 300 = 14700 & \overline{62224} \\
\text{And } 14700 \times 4 = & 58800 \\
\end{array}
$$

$$
\begin{array}{r}
405224 \\
\text{Proof } 74 \times 74 \times 74 = \quad 405224 \\
\end{array}
$$

And the cube root of $31\overset{\cdot}{1}89\overset{\cdot}{7}\cdot91\overset{\cdot}{0} = 67\cdot8$ nearly. Also the $\sqrt[3]{\cdot0001357} = \cdot05138$.

Applications of the Cube Root.

First. To find the proportion of solid bodies. *Rule.* All solid bodies are in proportion to each other as the cubes of their diameters, or similar sides.

Secondly. To find the side of a cube equal in solidity to any given solid. *Rule.* Extract the cube root of the solid content.

Thirdly. To find the dimensions of any solid body, having those of another given, either greater or less. *Rule.* Cube the dimensions respectively; enlarge or diminish them in the proportion required, and extract the cube root of the result.

We might increase these articles, but consider what has now been written sufficient for this work, in a subsequent part of which reference must be had to these applications of the square and cube root.

CHAP. II.

BOOK-KEEPING.

BOOK-KEEPING is the art of recording in a concise and systematic manner the transactions of all persons engaged in pursuits connected with money, in order that the amount of profit and loss in any business may be easily ascertained, and that the person keeping the books may at any time be enabled clearly to show the exact state of the money transactions that have passed through his hands. There are many systems of book-keeping, but they are all referrible to two kinds; one of which is termed keeping books by *single,* and the other by *double entry.* Of these single entry is by far the most simple, and it is most commonly applied to horticultural purposes; but the money transactions in a farm being more complicated, the books used by farmers are generally kept by double entry.

SECTION I.

GARDEN BOOK-KEEPING.

THE books kept by a gentleman's Gardener are generally very simple, and even in the largest establishments seldom exceed three in number; viz. the Cash-Book, the Labour-Book, and the Kitchen-Book; to which may be added a Day-Book, or Journal. Some Gardeners keep also a Cropping or Rotation Book, and an Order-Book; but these are so seldom used that it does not seem necessary to give examples of them.

The simplest form of book-keeping for Gardeners is where only one book is kept, which serves both as a Labour-Book and a Cash-Book.

E

This book should be of post-paper size, and ruled in the ordinary way of account books in general. On the outside should be written in a legible hand " GARDEN ACCOUNT," with the place, the year, and the day of the month that the account commences. The payments should be on the right-hand page, and the receipts on the left-hand page, as shown in the following example, forming folios 2. and 3. of the book:

(*Folio* 2.)

JANUARY, 1844.

		£	s.	d.
9	Received for 12 bushels of potatoes, at 2*s.* 9*d.* per bushel - - - - -	1	13	
13	Received for 15 bushels of potatoes, at 2*s.* 6*d.* per bushel, of Mr. John Chapman - -	1	17	6
	Carried forward - -	3	10	6

(*Folio* 3.)

JANUARY, 1844.

		£	s.	d.
	Brought forward - -	3	10	6
29	Received your draft to pay the balance of this account to this day - - -	7	10	6
	£	11	1	

Where a great number of work-people are employed it may be advisable to keep a separate book for labour, and so carry the amounts once a month, or at any fixed period, to the cash account just described. By this means the cost for labour would be known at one view, without being mixed up with occasional disbursements.

These separate Cash and Labour Books may be constructed on the same plan as the Cash and Labour Books

(*Folio 2.*)

JANUARY, 1844.

					£	s.	d.
8	John Smith, for	-	-	6 days' labour		12	
	Thomas Jones	-	-	2½ —		5	
	William Freeman	-	-	5 —		10	
	John Winchester	-	-	6 —		12	
	William Bates	-	-	4½ —		9	
15	John Smith -	-	-	5 —		10	
	Thomas Jones	-	-	6 —		12	
	William Freeman	-	-	6 —		12	
	John Winchester	-	-	6 —		12	
	William Bates	-	-	6 —		12	
	Paid for 12 birch brooms		-	-		4	
	Carried forward		-	-	5	10	

(*Folio 3.*)

JANUARY, 1844.

					£	s.	d.
	Brought forward		-	-	5	10	
22	John Smith	-	-	6 days' labour		12	
	Thomas Jones	-	-	6 —		12	
	William Freeman	-	-	6 —		12	
	John Winchester	-	-	5 —		10	
	William Bates	-	-	6 —		12	
29	John Smith	-	-	6 —		12	
	Thomas Jones	-	-	4½ —		9	
	William Freeman	-	-	6 —		12	
	John Winchester	-	-	6 —		12	
	William Bates	-	-	4 —		8	
				£	11	1	

recommended for foresters in p. 63, 64.; or only the Labour-Book may be on that model, and the Cash-Book may resemble that advised for nurserymen.   (See p. 70.)

An example of the mode of keeping a KITCHEN-BOOK is given in the following extract from the book kept by Mr. Barnes, gardener at Bicton, the seat of the Right Honourable Lady Rolle: —

Vegetable, Fruit, and Flower List, for the Week ending Saturday, Sept. 25. 1842.

	Sept.	19	20	21	22	23	24	25
Mushrooms	dish	1				1		
French Beans	,,	1	1	1	1		1	1
Warwick Peas	,,			1				
Early Frame Peas	,,					1		
Long-pod Beans	,,		1					
Windsor Beans	,,						1	
Cauliflowers	,,	1	1	1		1		1
Artichokes	,,				1			
Cape Broccoli	,,	1		1			1	1
Cabbage	,,	1	1	1	1	1	1	1
Greens or Coleworts	,,		1			1		
Turnips	,,	1	1	1	1	1	1	1
Carrots	,,	1	1	1	1	1	1	1
Potatoes	,,	1	1	1	1	1	1	1
Vegetable Marrow	,,		1				1	
Spinach	,,			1		1		
Silver Beet	,,	1						
Cucumbers for stewing	dish		3				5	
Peas	,,				1			
Lettuce	,,		1				1	
Endive	,,				1			
Red Cabbage	,,						1	
White Celery	,,		1			1		
Tomatoes	,,			1				
Horseradish	dish				1			1
Onions	,,	1	1	1	1	1	1	1
Shallots	,,		1				1	
Leeks	,,	1				1		
Garlic	,,			1				
Parsley, curled	bunch	1	1	1	1	1	1	1
Sweet Marjoram	,,		1				1	
Sweet Basil	,,			1			1	
Fennel	,,				1			
Tarragon	,,	1				1		
Green Mint	,,			1				
Chervil	,,	1	1		1	1	1	1
Sorrel	,,			1			1	

	Sept.	19	20	21	22	23	24	25
Winter Savory	bunch	1						
Chives	"					1		
Pennyroyal	"				1			

Salad for Servants.

		19	20	21	22	23	24	25
Cucumbers	dish	1		1		1		1
Lettuce	"	1	1	1	1	1	1	1
Radishes	"		1		1		1	

Picklings.

		19	20	21	22	23	24	25
Gherkin Cucumbers	-				200			
Onions, silver-skinned	peck			1				
Red Cabbage	doz.					1		
Capsicums	-			200				
Chillies	-					200		
Green Tomatoes	doz.		6					
Ripe Tomatoes for Sauce	"			7				

Plants and Flowers.

		19	20	21	22	23	24	25
Cut Flowers	basket	1			1			
Dahlias	doz.	4			4			
Magnolia Flowers	"	2			2			
Plants for Baskets in front hall	-	26			14			

Salad sent in for Table.

		19	20	21	22	23	24	25
Cucumbers	dish	1	1	1	1	1	1	1
Lettuce	"	1	1	1	1	1	1	1
Radishes	"	1	1	1	1	1	1	1
Celery	"		1		1	1		
Endive	"			1		1		
Red Beet	"	1						
Mustard and Cress	"		1				1	
American Cress	"				1			1

For Preserving.

		19	20	21	22	23	24	25
Orange Flowers	quart			10				
Magnolia Flowers	doz.			3				
Figs	"					2		
Grapes	basket			1				
Guava Fruit	doz.				8			
Damsons	quart			8				
Apples for Jelly	bush.					2		

Kitchen Fruit.

		19	20	21	22	23	24	25
Apples	peck	2			2			
Waste Peaches	-			9				

	Sept.	19	20	21	22	23	24	25
Plums - - -	quart					1		
Cherries - - -	lb.			1½				
Currants - - -	,,		1				1	
Raspberries - - -	,,		1				1	
Pears for stewing - -	doz.			2				
Apples for roasting -	,,					2		
Table Fruit.								
Pine-apples:								
Brown Sugar-loaf - -		1						
Queen - - - -					1			
Otaheite - - -							1	
Cýcas revolùta - - -		9				7		
Mùsa Cavendíshii - -			7		7		7	
Guava Fruit - - -				9		9		
Black Hamburg Grapes -	lb.	1	1	1	1	1	1	1
Sweetwater, Dutch . - -	,,	1		1		1		1
Muscat of Alexandria -	,,		1		1		1	
Peaches, Malta - - -		4			5		5	
Figs - - -	dish	1	1	1	1	1	1	1
Cherries - - -	,,		1		1		1	
Keen's Seedling Strawberries	,,	1		1	1		1	1
Red Currants - - -	,,	1	1	1	1	1	1	1
White ditto - - -	,,	1	1	1	1	1	1	1
Apples - - -	,,	1	1	1	1	1	1	1
Pears - - -	,,	1	1	1	1	1	1	1
Walnuts - - -	,,	1	1	1	1	1	1	1
Melons - - -	-	1	10	1		1		1
Impératrice Plums - -	dish				1	1	1	
Ice - - - -	-							

It will be observed that the gardens at Bicton are on a very large scale; and that consequently comparatively few gardens will need so extensive a Kitchen-Book as that from which the above example is extracted; it is, however, always useful to have some memorial of the produce sent in by the Gardener for the use of the house, as the Gardener is sometimes blamed for the non-productiveness of his department, while the fault, in fact, lies in the wastefulness or carelessness of the cook, or some other person belonging to the establishment. On this account, it is well for the gardener, if he does not keep a regular Kitchen-Book, to have, at least, a small Memorandum-Book in which he may enter on the left-hand side the things he sends to the kitchen, with the date, and the name of the person by whom he sends them; while on the right-hand side may be entered the receipt of the cook or other person to whom they are delivered. Sometimes a written paper is sent in with the things, which is receipted and returned by the cook; but as these pieces of paper are very liable to be lost, the Memorandum-Book will be found much more useful and equally convenient.

Section II.

BOOK-KEEPING FOR THE BAILIFF AND LAND-STEWARD.

The accounts of the Farm-Bailiff may be kept in one book, in a similar manner to those of the gardener, writing on the outside of the book the " FARM ACCOUNT," &c. &c. ; but if one person superintends both the garden and farm, in order to keep both establishments distinct, which should always be done, an additional book will be necessary, and the balances must be taken from the farm account, and carried to a general cash account, together with the garden account, and all other occasional receipts and payments.

Sometimes a Farm-Bailiff undertakes also the duties of a forester; and if he should have a large extent of land, with woods, &c., under his charge, he will necessarily have occasion both to give and take credit, proportioned to the extent of his dealings. His books should, therefore, be kept somewhat different to those before described. The fewer books he has, however, consistent with accuracy and explicitness, the better. An example is here given of a very simple method for registering all transactions of importance. The sales, it will be observed, are entered on the right-hand page, and the purchases on the left.

(*Folio 2.*)

JANUARY, 1844.

		£	s.	d.
7	Bought of Mr. W. Johnson 15 bushels of spring wheat for seed, at 9s. 6d. per bushel - Paid the 14th instant.	7	. 2	6
21	Bought this day of Mr. William Wiseman 50 store sheep; to be - - - Paid the 3d March.	75		
28	Bought on trial a four years' old cart horse (Ball), the price to be - - -	40		
	FEBRUARY.			
3	Bought of John Smith a cast-iron roller for the meadows, to be - - - -	10	10	

The title of this book should be "JOURNAL OF PECUNIARY TRANSACTIONS UPON THE ESTATE OF A. B., ESQUIRE;" and its size should be proportioned to last a year at least; observing that the quarto form is always the most convenient to take before a lady or gentleman for inspection. The book should have an alphabet at the beginning, and the names which occur in the book written against the letter corresponding with the initial of the surname, and the number affixed should show the page where such name will be found. Of course, when the names are entered, the surname should be first, as that is most likely to be looked for. It may be added, that all cash received and paid in this book should immediately be entered in the cash account, along with labour and all other incidental payments and receipts; and at the end of every year an

(*Folio 2.*)

JANUARY, 1844.

		£	s.	d.
3	Sold to Thomas Jones 40 Southdown sheep, to be paid for this day fortnight — the sum - Received the 17th.	80		
17	Sold to Mr. Buyer 3 loads of stack-wood, at 30s. Sold the same 500 fagots - - - Received 3d February.	4 5	10	
18	Sold John Hill, Esq., 6 loads of hay, at 75s. per load - - - - - Received the 2d February.	22	10	
29	Sold Mr. Brewer 10 quarters of barley, at 37s. 6d., to be paid for on delivery - - Received the 3d February.	18	15	
	FEBRUARY.			
5	Sold to Mr. Butcher a fat calf - -	4	5	

abstract should be carefully made of all the unsettled items, and entered at the beginning of the book for the following year.

When a Farm-Bailiff has a great many people upon particular works to superintend (and this will always be the case in the haymaking season), the following method of keeping their time may be adopted, and may be called the " HAY-MAKING ACCOUNT." This plan not only saves a great deal of time, but it becomes an interesting reference as compared with other years ; showing the commencement and duration of the haymaking, and consequently the comparative state of the weather for several successive years. It also shows the exact days each person has been at work or absent ; and, where the blanks occur, it may be inferred that those days were wet, or not fit for making hay.

HAY-MAKERS' TIME, 1844.

Began to mow the 8th of June. Names of Persons employed.	JUNE.																				JULY.				JULY.	£	s.	d.
	M 11	T 12	W 13	T 14	F 15	S 16	17	M 18	T 19	W 20	T 21	F 22	S 23	24	M 25	26	W 27	T 28	F 29	S 30	M 2	T 3	W 4				
John Smith	1	1	1	1	1			1	½	1	1	1	½		1		½	1	1	1	1	1	1	17½ days at 3s. 6d.	3	1	3
William Wilson	1	1	1	1	1	1		1	1	1	1	1	1		1		½	1	1	1	1	1	1	17 ,, 3s. 6d.	2	19	6
John Pitman	1	½	1	1	1	1		1	1	1	1	1	1		1		½	1	1	1	1	1	1	16¾ ,, 3s. 6d.	2	17	9
Thomas Stone				1	1	1		1	1	1	1	1			1		½	1	1	1	1	1	1	16½ ,, 2s. 6d.	2	1	3
William Roberts				1	1	1		1	½	1	1		1		1		½		1	1	1	1	1	15½ ,, 2s. 6d.	1	18	9
Jane Wilson								1	1		½				1		½	1		1	1		1	13 ,, 1s. 0d.	0	13	0
Mary Roberts				1	1	½		1	1	1	1	1	1		1		½	1	1	1		1	1	13½ ,, 1s. 0d.	0	13	6
Hannah Smith				1	1	1		1	1	1	1	1	1		1		½	1	1	1	1	1	1	15½ ,, 1s. 0d.	0	15	6
Ann Hester				1		1		1	1	1	1	1	1		1		½	1	1	1	1	1	1	15½ ,, 1s. 0d.	0	15	6
John Wilson				1		1		1	1	1	1	1	1		1		½	1	1	1	1	1	1	14½ ,, 3s. 0d.	2	3	6
						Sunday							Sunday		Rainy day						Sunday				17	19	6

The above example it is hoped will need but little explanation, for its utility to be appreciated. It is extremely useful in very heavy jobs where a number of irregular hands have been called in, and particularly so when their employment had been varied from piece-work to time-work, such as road-making, &c. When once the name is written it only in future requires filling up with the figure 1 for the whole day, $\frac{1}{2}$ for half a day, and $\frac{1}{4}$ for the quarter day. Moreover, as the precise time can at once be referred to, there is no danger of mistakes when the account runs on, which is not unfrequently the case.

In addition to the preceding books, the Farm-Bailiff, in extensive establishments, may keep separate or additional accounts for the expenses of coach and saddle horses wanted by the family at the mansion, and probably of a poultry-yard, an aviary, or a menagerie.

The Land-Steward requires a Cash-Book, a Day-Book, and a Ledger.

Professor Donaldson also recommends a Rotation-Book,

ROTATION-BOOK.
Years of Lease, 14.

FIELDS.	1844.	1845.	1846.	1847.	1848.	1849.	1850.	&c. &c.
No. 1	Meadow — when dunged.							
2	Tur.	Bar.						
3	Bar.	Clov.						
4	Clov.	Oats.						
5	Oats.	Tur.						
6	Bar.	Clov.						
7	Oats.	Tur.						
8	Clov.	Oats.						
9	Tur.	Bar.						
10	Wht.	Vet.						
11	Oats.	Fall.						
12	Vet.	Wht.						
13	Fall.	Wht.						
14	Clov.	Oats.						

The above example should exhibit fourteen columns, the years of the lease, but the breadth of our page will not admit of so many.

and that " every estate should be provided with a map on
a large scale; and also with a book of maps, on which
to mark any improvements suggested. From the book of
maps, a plan of each farm separately, on a scale suited for
a large pocket-book, may be made in the Rotation-Book,
on an opposite page to that containing a list of the successive crops, and with the name of the farm attached;
forming a convenient book of maps for field consultation.
The rotation may be extended to any number of years,
and may be made to suit any course of cropping. On
small estates, the agent, or person in charge of it, will
examine each field on every farm twice in the year; in
the spring, and in the end of autumn, and mark in the
proper column in the Rotation-Book, the kind of crop the
field has carried. On extensive concerns, this duty will
devolve on the Manor-Bailiff, who must reside on the property. By means of this examination of the field, any
deviation from the prescribed or good husbandry is imme-

CASH-

(*Folio* 2.)   ROBERT STEWARD *in Account*

		£	s.	d.
	Dʀ.	£	s.	d.
1843	Brought forward - -	550	19	9
June 10	Received of J. Jones, Esq., for two years'			
	quit rent - - - -	5	3	4
15	Received of Mr. John Knight, an arrear			
	of rent, due Lady-day last - -	15		
Nov. 8	Received at the audit, held this day at			
	the George Inn, Auborn - -	1573	10	9

In the Ledger an account is opened for every farm, cottage, or
separate holding on the estate. Where there are a number of ma-

diately detected; and also if the yearly cultivation of the field be properly performed. Such a check would prevent quarrels and litigations about damages at the end of leases, on account of bad and scourging farming; and any land-owner, by examining his Rotation-Book at home, is able to see the cropping of the whole estate without the trouble of inspection. Attention on these points would do much to introduce better cultivation; but great carelessness is almost everywhere evinced, very few proprietors either know or employ qualified men, and in no other art practised in our day has a recommendation to office ever proceeded from a total ignorance of the art itself, and from being engaged in a profession totally and irreconcilably different. The consequences are quickly and amply evident, and must be certain."

The following are examples of a Cash-Book and Ledger, taken from the books of the Land-Steward of an extensive estate :—

BOOK.

with JOHN PAYMENT, *Esq.* (*Folio* 2.)

		CR.	£	s.	d.
1843		Brought forward - -	340	13	6
Nov. 8		Allowed as agreed upon to Mr. John Smith for improving his house - -	20		
	8	Paid John Hall, the Bricklayer, as per bill, for repairing Mr. Nason's house -	9	7	6
	8	Paid for the audit dinner, as per bill, held at the " George," Auburn - -	15	15	
	8	Paid for stamps to receipt the rents paid at the audit - - -	3	12	
	8	Paid back to the tenants the Income Tax, amounting to - - -	45	17	
	10	Paid to your account at the bank of Messrs. Coutts and Co. - - -	1100		
		Paid the labourers' account, for the five weeks ending October 28th - -	75		

norial and other rights to be kept up, other accounts require to be opened, which each particular case will readily point out.

LEDGER.

RENTAL OF THE AUBORN ESTATE. Michaelmas, 1843.

	Tenants' Names.	Yearly Rents. £ s. d.	Arrears due, unpaid Lady-day 1843. £ s. d.	Half-yearly Rents. £ s. d.	Arrears due Michaelmas 1843. £ s. d.	Income Tax for Half a Year to Michaelmas 1843. £ s. d.	Amounts due less the Income Tax, Michaelmas 1843. £ s. d.	Cash received at the Audit, November 8th, 1843. £ s. d.	Arrears due and unpaid at the Audit, November 8th, 1843. £ s. d.	Remarks.
1	John Smith -	10 0 0		5 0 0	5 0 0	0 2 11	4 17 1	4 17 1		
2	Rev. Robert Holt -	19 10 0		9 15 0	9 15 0	0 5 6½	9 9 5½	9 9 5½		
3	John Atkins -	26 0 0		13 0 0	13 0 0	0 7 7	12 12 5	12 12 5		
4	Thomas Jones -	15 15 0	0 8 1 3	7 17 6	15 18 9	0 4 4½	15 14 4½	8 0 0	7 14 4½	Some timber promised for repairs.
5	William Hester, Esq. -	12 0 0	0 4 10 6	6 0 0	0 10 10 6	0 3 6	10 7 0	6 0 0	4 7 0	

Section III.

BOOK-KEEPING FOR THE FORESTER.

On an estate where the woods and plantations are extensive, the Forester will be required to keep a Labour or Time
Book, a Cash-Book, a Day-Book, and a Timber Sale-Book,
besides the Pocket Memorandum-Book, which is required
to assist the memory in every description of business.

The following models of these books have been kindly
supplied to us by the Forester of one of the best managed
wooded estates in England : —

LABOUR-BOOK.

1844.		Monday.	Tuesday.	Wednesday.	Thursday.	Friday.	Saturday.	Days.	Rate.	£	s.	d.
June 12	John Watson	1	1	½	0	1	1	4½	2s.		9	0
	Thomas Johnson	1	1	1	1	1	1	6	2s.		12	0
	Edward Johnson and Company 30 poles of hedge work at 2s. 6d.									3	15	0
26	Entered in the cash-book of this date									4	16	0

" Payment may be made every Saturday. I pay once
a fortnight, and on Monday, as it tends to keep the worst
characters quiet on the Sabbath day."

[This is an excellent plan; but in such localities as have
a market town adjacent to, or within reach of them, it
would, we think, be well for the Forester, and, indeed,
every one whose office it is to pay the labourers, to take
into consideration the advantage their workmen would
derive by being enabled to procure their provisions at the
market; and accordingly fix the day previous to that on
which the market is held, for payment.]

CASH-BOOK.

1844.			Debtor.			Creditor.		
			£	s.	d.	£	s.	d.
Sept.	16	To Mr. A. B., Leadenhall Street, in payment of 40 loads of bark - -	720					
		By Mr. A. B., commission 2½ per cent on the above -				18		
	18	By Mr. S. S. (the steward), remitted him - -				650		
	20	To proceeds of auction sale of June 10th - -	320	10	0			
		By amount of above, remitted Mr. S. S. - - - -				320	10	0
		To Mr. C. B., lots 12, 14, 23, 36, in New Hay Coppice sale, as per day-book of August 24. - -	20	15	0			
			1061	5	0	988	10	0
			988	10	0			
		Balance at this date - -	72	15	0			
		By labour-book of this date -				60	15	0

" *Remarks.* I balance with the steward every half year; but when I have large receipts I remit to him the greater part of the amount, only reserving a few pounds to pay my men with; and when I fall short I get what I want of him on giving him a receipt for what I get. I prefer this mode of settling, because I do not think it is quite right for inferior servants to keep large sums of cash in hand, and a master is sure to like one all the better for it, whatever be the amount of confidence he places in his servant."

DAY-BOOK.

June 16th 1844.	£	s.	d.
Mr A. B., Leadenhall Street, London, 40 loads of hatched bark, at 18*l*. - - -	720	0	0
August 24th.			
Mr. C. D. of Warrington, Lots 12, 14, 23, 36, in New Hay Coppice sale - - -	20	15	0

" *Remarks.*

" Parties conducting a sale of bark may either buy or
sell by the load, which is 2¼ tons, or by the ton of 20 cwt.
When bark lies in the woods and fit to carry to be stored,
if sold in this state, the tanner generally has it by what is
called long weight, that is, 120 lbs. instead of 112 lbs.
This is an allowance for shrinking, as the bark becomes
drier.   In selling the bark of large trees, this allowance
will come very near the mark; but in young coppice
bark, I think the allowance too much.   Much depends
upon the state in which the bark is in with regard to dry-
ness.   If the tanner, as is sometimes the case, shall be at
the expense of stripping off the bark, of course this expense is taken into consideration when the bargain is made,
and he is certain to take care to let it lie out as long as is
safe to do so, for the purpose of having it so dry as to prevent much shrinking.   The expense of peeling bark from
old trees in most places is 16s. per ton, after it is dry, and
the seller is at the expense of falling the trees.   The trees
are falled at different prices, according to their size, 2s. to
3s. and upwards, each; or, where there are a number,
they may be lumped.   Young coppice bark like mine requires more money in consequence of the smallness of the
trees.   I allow 30s. for peeling, falling, loading on the
waggons when fit to carry, and building in ricks, or storing
in the barn.   In some parts of the country, the woodman
goes round, and depending on his judgment, supposes that
so many trees are worth so much money, and the farmer
falls the trees, and peels at his own cost, and pays so much
the pound's value, sometimes 6s. 8d. per pound, and 11s.
has, at some times, been given.   This system of selling is
liable to serious objections.   No proprietor ought to submit
to it; for, if, at the conclusion of the season, the tanner
comes out with a good dinner to the woodman and a present, the question is, whether that woodman with such a
temptation before him, would in future value very high.
Some woodmen sell their bark at so much for a foot of

timber in a tree. This I think a confused way of doing
it, for a tree may contain 100 feet of timber, and still
another, containing only 50 feet, may produce more bark,
in consequence of having a large spreading head. Depend
upon it there is no mode of selling bark so satisfactory and
fair to all parties as that of selling by the ton or load,
either hatched or unhatched. I omitted to say, that in
Oxfordshire and other places, bark is hatched at 1*l.* per
load of 45 cwt. I give 10*s.* per ton."

Timber Sale-Book, 1844.

Lot.	Contents.	Value.			June.	Sold.			Reserved.		
No.		£	s.	d.		£	s.	d.	£	s.	d.
					Timber and Lop in —— Coppice.						
1	6 oak	10	0		100 feet at 2*s.*, Mr. A. B. -	10					
2	12 ash	8	15		70 feet at 2*s.* 6*d.*, Mr. B. C. -	8	15				
3	100		12		Fagots, Mr. C. D. - -		12				
4	Lop	1	4		1 Stack, Mr. C. D. - -	1	4				
5	Lop		12		½ Stack used for charcoal -					12	
6	300	1	16		Fagots, gift to Stablemen -				1	16	
7	10 elm	3	15		50 feet at 1*s.* 6*d.*, home use -				3	15	
						20	11		6	3	

" *Remarks.*

" When the woodman puts his valuation on any number
of lots, whether the sale is by private contract, or by auction, it may be that his customers will not come up to his
price; but he must in this case get as much as he can, and
enter the receipt in the sold column ; or, if he cannot get a
fair price, it must go to home purposes. If the sale is by
auction, and there are 100 lots to be disposed of, enter the
whole number, from first to last, in the book, with the
name of each purchaser, and the price. Every thing
should be lotted and numbered, unless it is a single tree or
two ; and in this case, such tree or trees should be entered
with the name of the buyer, and the name of the close or
coppice in which they grew."

Before proceeding to give the forms of books recommended to be kept by those conducting business on their own account, we would here most strongly advise all young men to be very particular in demanding vouchers for every payment they make, however small; and if they had a book in which to paste every voucher they receive, they would find it very convenient. A book of this kind should have an index at the beginning, referring to the page where each receipt is affixed. Indeed, we cannot impress too strongly upon the minds of young men engaged in places of trust, the necessity there is for accuracy and regularity in all their pecuniary transactions. It is not enough to be conscious of their own integrity, but they should always be prepared, when called upon, to prove their rectitude.

SECTION IV.

BOOK-KEEPING FOR THE NURSERYMAN.

As young gardeners sometimes become nurserymen, it may be useful to show the mode of book-keeping generally employed in such establishments; and the following models have been kindly sent us by the proprietor of a first-rate nursery establishment : —

The books required are the following : namely, Ledger, Journal, Waste-Book, Bought-Book, Labour-Book, Cash-Book, and Petty Cash-Book.

The four first-named books have alphabets with an index referring to the page where each person's name is entered, and the following copies from a page in each book will show their use. But the short description subjoined will make them still better understood.

The LEDGER is the book into which all accounts are posted, debiting the parties for goods supplied by us, and crediting them for cash or goods consigned to us.

(*Folio 2.*)   LEDGER.

DR.   THE RT. HONOURABLE LORD

1842.				J. C.	£	s.	d.
Sept.	20	To Goods -	- -	333	3	16	
1843.							
Feb.	8	To Do. -	- -	434	16	2	4½
					19	18	4½

The column J. C. refers to the page in the Journal C.

The JOURNAL has entered into it an exact copy of all invoices sent by post, &c.; and the amounts, with the date of the transaction, from time to time carried to the Ledger.

JOURNAL.

Thursday, October 26th, 1843.			
SIR JOHN L———, BART. H—— E——.	£	s.	d.
1 *Pí*cea nóbilis - - -	1	11	6
1 *Pì*nus Lambert*iàna* - - -	1	1	
24 do. Pallas*iàna* - - -	1	16	
2 do. *Pí*nea - - - -		5	
1 Basket by the down carrier.	4	13	6

28th.			
REV. JAMES T———, C———, SOMERSET.	£	s.	d.
1 Paul*ò*wn*ia* imperiàlis - -		5	
1 *A*`bies Menzi*èsii* - - -		7	6
1 *Pì*nus *Teoc*ò*te* 7s. 6d. Basket, &c. 1s. 6d. -		9	
By Great Western Railway.	1	1	6

LEDGER. (*Folio 2.*)

R——, B—— PARK. CR.

1843.				C.B.	£	s.	d.
August	24	By Cash - - -		81	19	18	4½

The column c. b. refers to the page in the Cash-Book.

The WASTE-BOOK is a sort of day ledger, in which entries are made of goods supplied near home, and of frequent occurrence. A portion of this book is set apart for each name, and the amounts at the time the bills are sent in are entered in the Ledger.

WASTE-BOOK.

HIS GRACE THE DUKE OF B——

1843.			£	s.	d.
Sept.	28	6 Magnòlia glaúca - - -	1	5	
		2 Andrómeda floribúnda - -		15	
		2 Kálmia latifòlia - - -		5	
Oct.	11	1 Jasmìnum azóricum - - -		1	6
		1 Pernéttia mucronàta - -		1	6
		1 Jasmìnum Wallichiànum - -		1	6

The BOUGHT-BOOK has, in the manner of the Waste-Book, a portion of it set apart for the name of each creditor, with an index· referring to the page, and the amount carried periodically to the creditor side of the Ledger.

BOUGHT-BOOK.

MESSRS. V—— AND SON, NURSERYMEN.

1843.			£	s.	d.
June	8	1 Echìtes spléndens - - -	5	5	
		1 do. atropurpùrea - -	2	2	
		1 do. hirsùta - - -	1	1	

The LABOUR-BOOK is simply an account of the time and wages paid to persons employed in the establishment, and is carried weekly to the cash account.

The CASH-BOOK has entered into it all moneys paid and received, with columns for figures referring to the page in the Ledger, Journal, Waste-Book, or Bought-Book where the particulars of the transaction are detailed.

The PETTY CASH is for small sums received in ready money, and for small payments, such as tolls, booking, letters, stamps, &c. &c., and the balances carried periodically to the Cash-Book.

SECTION V.

FARM BOOK-KEEPING.

THE correct and daily registration of matters relating to accounts is of the greatest importance to every person concerned in the transactions that result in gain or loss. Hitherto the consideration of the absolute necessity of a correct and simple method of Book-keeping has been mainly confined to the merchant and trader; and almost all the systems that have been established and are in use, have been formed with a view to their application to commercial accounts. The increasing importance, however, that is now given to the value of ascertaining correctly the results arising from the different modes of cultivating and stocking farms, rearing cattle, &c., makes it compulsory on those who are engaged in farming occupations to be in no way behind the commercialist in the knowledge of keeping accounts. The want of an uniform method for agricultural accounts has been very generally admitted, and it is for the purpose of supplying this defect that the accompanying system has been prepared. The books that are necessary to the carrying out of this plan are : —.

1st, Day-Book.
2d, Invoice-Book.
3d, Labour-Book.
4th, Cash-Book.
5th, Ledger.

To these may be added certain subsidiary books, such as Petty Cash-Book, Bill-Book, House and Personal Expense-Book, Letter-Book, Diary of Labourers' Work, &c. &c.; but as these are mere books of convenience, either to abridge the entries in any of the principal books, or for some other purpose, and can be accommodated, without any difficulty, to this system, it is deemed unnecessary to particularise them.

The principle of double entry has been applied to the accompanying system; that is, for every sum carried to the debit of any account, there is a corresponding amount placed to the credit of another account. Thus: if I buy goods of A, the account representing those goods in the Ledger is *debited*, and A. credited; if I pay A. 100*l.*, A. is debited, and the Cash-Book credited.

Subsect. I. — THE DAY-BOOK.

This book is intended to contain a register of all sales that are effected, entered under the date at which the occurrence takes place. The entry should be stated in the fullest and most complete manner, and should be intelligible to every one. The accompanying form shows the method in which the entries should be made; and in order to obviate the necessity of keeping a Journal, as in use among merchants, a double folio column is introduced for the purpose of posting each item; in the first place, to the debit of the personal account, and in the next, to the credit of the nominal account.

DAY-BOOK.

1845.				Cr. Fol.	Dr. Fol.	£	s.	d.	£	s.	d.
Jan.	16	RICHARD SEWELL, of Highbeach.		L							
		1½ rick of hay, at per load 60s.		1		30	0	0			
		17 pigs - - each 10s.		4		8	10	0			
		1 rick of wheat - -		1	L	58	0	0			
					5				96	10	0
	19	HARVEY COMBE.									
		40 loads of bark - at 18l.		1		720	0	0			
					5				720	0	0
	21	ISAAC MARCH.									
		1 horse (Blackleg) - -		3		25	0	0			
		1 cow - - -		3		12	0	0			
		2 oxen - - each 16l.				32	0	0			
		3 ewes - - „ 50s.		3		7	10	0			
		6 lambs - - „ 8s.				2	8	0			
					5				78	18	0

The entry of sale should be made as soon as the sale is
effected, and posted to the debit in the Ledger, of the party
purchasing; the folio of his account in the Ledger being
placed in " debtor folio," while the account opened for
each description of goods in the Ledger must be credited
for the like amount, and the folio placed in the " credit
folio."

Subsect. II.—THE INVOICE-BOOK.

The Invoice-Book is intended to contain a register of
all purchases made; and the mode of keeping this book is
more fully explained in the form which we have given.
The double folio column is also introduced in this, for the
same purpose as stated in Subsect. I., with this difference,
that the personal account is credited, and the nominal
accounts are debited, with each item.

INVOICE OF PURCHASES.

Date.			Fol. Dr.	Fol. Cr.	£	s.	d.
1845.		This book should be simply a guard-book, made of common blue paper; and the invoice of goods purchased should be neatly folded into a small compass, and pasted into this space; the top edge only being pasted down, the invoice can at any time be unfolded for the purpose of ascertaining its contents. On the outside should be endorsed the name thus :					
Jan.	16	" HENRY SMITH," and the date and amount entered against it, as here shown. The purchases are to be posted to the credit of the party of whom they are made, and the debit of the account opened for the goods purchased.	4	6	72	14	3

The above example is intended to represent a purchase of farming implements; and on reference to folio 6. it is seen that the party who sold the goods has credit, and at folio 4. implements, &c., are debited, for the amount of the invoice.

___

Subsect. III.—THE LABOUR-BOOK.

The Labour-Book is formed so as to be a register of the time any servant is engaged on each day, and the total amount due to each at the end of the week is carried into the outer column. This book should be added up every week, and the total amount paid for wages carried to the credit of the Cash-Book.

This book may be so formed as to enable the farmer to state upon what work the labourer has been employed; but if so kept, it would be necessary, to carry it out fully, to open accounts in the Ledger for each portion of the farm under different cultivation, charging each with the expenses of cultivation, cost of seed, &c.

LABOUR-BOOK.

The Week ending the 4th Day of January, 1845.

Names.	Monday.	Tuesday.	Wednesday.	Thursday.	Friday.	Saturday.	Total Days.	Rate per Diem.	£	s.	d.
Thomas Stevens	1	1	$\frac{1}{2}$	$\frac{3}{4}$	1	1	$5\frac{1}{4}$	s. d. 2 6	0	13	$1\frac{1}{4}$
							C.B. 2		0	13	$1\frac{1}{4}$

Note. The weekly total is entered in the Cash-Book, with a reference to the Labour-Book folio against the entry.

Subsect. IV. — THE CASH-BOOK.

The Cash-Book is debited with all sums received, and credited with all moneys paid away; and these entries should be made at the time the transaction takes place, in order to prevent confusion and mistakes. The balance should be struck once a month, and it should agree with the cash on hand. At short intervals the entries should be posted into the Ledger.

CASH-BOOK.

(Folio 1.) Dr.

CASH.

1845. Jan.			L	£	s.	d.
1	To balance in hand, On taking possession of Oakley farm -		7	1720	5	0
17	To Richard Sewell. Received of him for sales of hay, &c. -		5	96	10	0
21	To Harvey Combe. Received of him for bark		5	720	0	0
25	To Isaac March. Received of him for horse -		5	25	0	0
27	To Smith and Co. Received of them -		7	509	8	5

(Folio 2.) Cr.

PER CONTRA.

1845. Jan.			L	£	s.	d.
1	By Richard Walker. Paid him on taking possession of Oakley farm in part for stock, &c., as per account rendered at that time -		6	1427	16	7
25	By Henry Smith. Paid him for implements		6	72	14	3
	By Smith and Co. (Bankers.) Paid to my credit with them - - -		7	900	0	0
	By wages. Week ending 4th day of January	L.B 1	2	0	13	1½
27	By R. Walker. Paid him on account of valuation of property at Oakley farm -		6	509	8	5

Subsect. V. — The Ledger.

The Ledger contains the result of all the transactions
that are entered in the other books; and for that purpose
is divided into numerous accounts under proper titles,
according to the articles they are intended to represent.

The three distinctive heads of accounts are called Personal, Real, and Nominal.

1. *Personal accounts* are those which represent transactions with individuals. The debit side of these accounts
exhibits the amount due from any party in respect to goods
sold, or cash paid; the credit side the amount of cash received or goods purchased. The balance of these accounts
shows, if the debit side is the larger, the amount due from,
if the credit side is the larger, the amount due to, the party.

2. *Real accounts* are those which represent property,
such as stock in trade, freeholds and leaseholds, shares in
companies, implements in trade, &c. These accounts are
debited with the value of the respective assets on hand,
and with all sums paid for the further purchases of a like
description; or, of payments made or goods purchased in
respect of such property, and credited with cash received
from the sale of each property. Particular care should be
taken that receipts, such as rents, profits on the trading of
vessels, or any similar receipts that do not diminish the
value of the property, be not posted to the credit of the
account representing the cost or value of the property. If
the property be of such a character as though yielding an
income, yet, as in the usual case of leaseholds, becomes
yearly of less value, then such accounts must be periodically debited with interest equivalent to the estimated
amount of the diminution in value in such period.

3. *Nominal accounts* are those which are rendered necessary for the purpose of accurately ascertaining the results of trading, and comprise Profit and Loss, Interest
and Discount, Charges in Trade, Wages and Expenses,
House and Personal Expenses.

All these accounts contain, on the debit side, the moneys
paid, or other items which are a charge upon the con-

ducting of the business, and are credited with any amounts which are matters of gain on any of the accounts.

It may be useful to specify more in detail the uses of the account of Profit and Loss, into which all the other accounts are eventually merged.

Profit and Loss. The accounts in the accompanying Ledger, representing the various property, having been duly made up by being debited with the value of stock, &c., on the one hand, and charged with all further payments made in respect of any of them, and credited with all the sales effected, must then be credited with the quantity and value of stock on hand at the time of balancing the books, and the difference, if the debit side is greater, must be carried to the debit, or if the credit side be greater, to the credit of Profit and Loss, and the stock on hand brought down in each account as a debit balance. The transfers from these accounts being then passed to the Profit and Loss, the gross gain or loss is shown upon the transactions of the different commodities traded in; but as the gain has to be diminished, or the loss increased, by the expenses attending the business, the total of the wages, trade charges, interests, &c., must be transferred to the debit of this account in order to ascertain the net loss or profit on the year's transactions. The balance of this account is then transferred to the capital account, which latter account is debited with the house and personal expenses.

Having thus defined, as clearly as the limits of this section will admit, the mode of opening, continuing, and balancing the various accounts, it only remains to explain the test adopted for ascertaining the accuracy of the work. This consists in taking out all the balances from the Ledger under the respective heads of debit and credit, together with the balance of the Cash-Book. This is termed a "trial balance," and, when added up, should be the same on both sides; a disagreement between the two being a proof of an error somewhere; or, in other words, that every debit has not a corresponding credit, which method should invariably be maintained throughout the accounts.

FARMER'S LEDGER.

INDEX.

N.	O.

P.

Petty Charges   -   -   2
Poultry-Yard   -   -   4
Profit and Loss on Live Stock   2

Q.

R.

Rick Yard   -   -   1

S.

Sheep   -   -   3
Swine   -   -   4
Sewell, Richard   -   5
Smith, Henry   -   6
Smith & Co.   -   7

T.

U, V.

W.

Wages and Expenses of cultivation   -   -   -   2
Walker, Richard   -   -   6

X.

Y.

Z.

(1)

| Dr. | | | | | | | | | | | RICK- | | |

Date. 1845.		Folio.	Wheat.	Oats.	Rye.	Beans.	Barley.	Hay.	Peas.	£	s.	d.
Jan. 1	To stock purchased of R. Walker.	L 6	4	3	6	2	5	7	2	510	0	0

| Dr. | | | | | | | | | | | | | BARNS. | | |

Date. 1845.		Folio.	Wheat. Q. B.	Oats. Q. B.	Rye. Q. B.	Beans. Q. B.	Barley. Q. B.	Peas. Q. B.	Sundries.	£	s.	d.
Jan. 1	To Stock purchased of R. Walker. Bark - -	L 6	- -	- -	- -	- -	- -	- -	Lds. 100	1400	0	0

| Dr. | | | | | | | | | | | | | GRANARY. | | |

Date.		Folio.	Wheat. Q. B.	Oats. Q. B.	Rye. Q. B.	Beans. Q. B.	Barley. Q. B.	Peas. Q. B.	Sundries.	£	s.	d.

(1)

YARD.											Cr.		
Date. 1845.		Folio.	Wheat.	Oats.	Rye.	Beans.	Barley.	Hay.	Peas.	£	s.	d.	
Jan. 16	By R. Sewell -	D. B. 1	-	-	-	-	-	1½	-	30	0	0	
	By ditto -	1	1	-	-	-	-	-	-	58	0	0	

																			Cr.		
Date. 1845.		Folio.	Wheat.		Oats.		Rye.		Beans.		Barley.		Peas.		Sundries.		£	s.	d.
			Q.	B.	Q.	B.	Q.	B.	Q.	B.	Q.	B.	Q.	B.					
Jan. 19	By H. Combe. Bark -	D. B. 1	-	-	-	-	-	-	-	-	-	-	.	-	Lds. 40	720	0	0

																			Cr.		
Date.		Folio.	Wheat.		Oats.		Rye.		Beans.		Barley.		Peas.		Sundries.		£	s.	d.
			Q.	B.	Q.	B.	Q.	B.	Q.	B.	Q.	B.	Q.	B.					

(2)

Dr.					WAGES AND EXPENSES			
1845. Jan.	4	To Cash	-	-	-	-	C.B. 2	£	*s.* 13	*d.* 1½

Dr.								PETTY

Dr.						PROFIT AND LOSS		

(2)

OF CULTIVATION. Cr.

CHARGES. Cr.

ON LIVE STOCK. Cr.

(3)

| Dr. | | | | | | | | | CATTLE. | | |

			Folio.	Bulls.	Oxen.	Cows.	Heifers.	Calves.	£	s.	d.
1845. Jan.	1	To R. Walker, for amount included in the valuation -	L 6	2	-	20	10	-	290	0	0

| Dr. | | | | | | | | HORSES. | | |

			Folio.	Horses.	Mares.	Foals.	£	s.	d.
1845. Jan.	1	To R. Walker, for amount in- cluded in the valuation -	L 6	12	-	-	180	0	0

| Dr. | | | | | | | | | SHEEP. | | |

			Folio.	Tups.	Ewes.	Shears.	Hogs.	Lambs.	£	s.	d.
1845. Jan.	1	To R. Walker, for amount included in the valuation -	L 6	-	200	-	-	170	268	0	0

(3)

			Folio.	Bulls.	Oxen.	Cows.	Heifers.	Calves.	£	s.	d.
											Cr.
1845. Jan.	21	By Isaac March -	D.B. 1	-	2	1	-	-	44	0	0

			Folio.	Horses.	Mares.	Foals.	£	s.	d.
									Cr.
1845. Jan.	21	By Isaac March - - -	D.B. 1	1	-	-	25	0	0

			Folio.	Tups.	Ewes.	Shears.	Hogs.	Lambs.	£	s.	d.
											Cr.
1845. Jan.	21	By Isaac March -	D.B. 1	-	3	-	-	6	9	18	0

(4)

Dr.					Folio.	Boars.	Sows.	Pigs.	£	s.	d.	SWINE.
1845. Jan.	1	To R. Walker, for amount included in the valuation -			L 6	-	-	19	14	5	0	

Dr.				Folio.	Butter.	Cheese.	Milk.	Eggs.	Sundries.	£	s.	d.	DAIRY AND POULTRY

Dr.				£	s.	d.	IMPLEMENTS, CARTS,
1845. Jan.	1	To R. Walker, for amount included in the valuation - - - -	L 6	275	0	0	
	16	To H. Smith - - - -	I.B. 1	72	14	3	

(4)

					Folio.	Boars.	Sows.	Pigs.	£	s.	d.
				Cr.							
1845. Jan.	16	By Richard Sewell	-	-	D.B. 1	-	-	17	8	10	0

YARD. **Cr.**

				Folio	Butter.	Cheese.	Milk.	Eggs.	Sundries.	£	s.	d.

HARNESS, &c. &c. **Cr.**

			£	s.	d.

(5)

Dr.						RICHARD		
1845. January	16	To Goods - - -		D.B. 1	£ 96	s. 10	d. 0

Dr.						HARVEY		
1845. January	19	To Goods - - -		D.B. 1	£ 720	s. 0	d. 0

Dr.						ISAAC		
1845. January	21	To Goods - - -		D.B. 1	£ 78	s. 18	d. 0
		To Balance brought down - -		-	53	18	0

(5)

SEWELL, of Highbeach.					Cr.		
1845. January	17	By Cash - - -	C. B. 1	£ 96	s. 10	d. 0	

COMBE.					Cr.		
1845. January	21	By Cash -	C. B. 1	£ 720	s. 0	d. 0	

MARCH.					Cr.		
1845. January	25	By Cash - - -	C. B. 1	£ 25	s. 0	d. 0
		By Balance carried down - -	-	53	18	0
				78	18	0

(6)

Dr.					HENRY		
1845. Jan.	25	To Cash - - -	C. B. 2	£ 72	s. 14	d. 3	

Dr.					RICHARD		
1845. Jan.	1	To Cash - - -	C. B. 2	£ 1427	s. 16	d. 7
	27	To „ - - -	2	509	8	5
		To Balance carried down -	-	1000	0	0
				2937	5	0

(6)

SMITH.						Cr.		
1845. Jan.	16	By Goods - - -			I. B. 1	£ 72	s. 14	d. 3

WALKER.						Cr.		
1845. Jan.	1	By the following stock purchased of him at the time of entering upon the Oakley farm, as per valuation. By				£	s.	d.
		Dead Stock. £ s. d. 4 ricks of wheat - 100 0 0 3 „ oats - - 60 0 0 6 „ rye - - 90 0 0 2 „ beans - - 30 0 0 5 „ barley - - 100 0 0 7 „ hay - - 100 0 0 2 „ peas - - 30 0 0			L 1	510	0	0
		(As per Rick-Yard Account, fo. 1.) 100 loads of bark - 1400 0 0			1	1400	0	0
		Cattle. £ s. d. 2 bulls - - 40 0 0 20 cows - - 200 0 0 10 heifers - - 50 0 0			3	290	0	0
		Sheep. £ s. d. 200 ewes - - 200 0 0 170 lambs - - 68 0 0			3	268	0	0
		Horses. — 12 horses - -			3	180	0	0
		Swine. — 19 pigs - -			4	14	5	0
		Implements, carts, harness, &c., as per inventory - - -			4	275	0	0
						2937	5	0
		By balance brought down - -			-	1000	0	0

(7)

Dr.					SMITH & Co.,			
1845. January	25	To Cash paid them	-	-	C. B. 2	£ 900	s. 0	d. 0

Dr.				CAPITAL.		
				£	s.	d.

(7)

Bankers.						Cr.		
1845. January	27	By Cash	-	-	C.B. 1	£ 509	s. 8	d. 5

						Cr.		
1845. January	1	By Cash at this date	-	-	C.B. 1	£ 1720	s. 5	d. 0

TRIAL BALANCE.

Folio. c.b.		Dr.			Cr.		
		£	s.	d.	£	s.	d.
1	Cash - - - -	160	11	0½			
L. 1	Rick-Yard - - -	422	0	0			
	Barns - - - -	680	0	0			
2	Wages and Expenses of Cultivation		13	1½			
3	Cattle - - -	246	0	0			
	Horses - - -	155	0	0			
	Sheep - - -	258	2	0			
4	Swine - - -	5	15	0			
	Implements, Carts, &c. -	347	14	3			
5	Isaac March - - -	53	18	0			
6	R. Walker - - -				1000	0	0
7	Smith & Co. - - -	390	11	7			
	Capital - - -				1720	5	0
		2720	5	0	2720	5	0

CHAP. III.
PRACTICAL GEOMETRY.

Section I.
OF THE SCALE AND COMPASSES.

PRESUMING that the reader has made himself master of the previous part of this work, which embraces the science of numbers, he will now enter upon a branch of elementary mathematics, which will prove both interesting and instructive to him. To this end he must provide himself with a *Scale* and *Compasses.* The scales are made of hard wood or ivory, and usually with a variety of lines drawn upon them, such as scales of chords, &c. &c.; but the *scale of equal parts* is that which we have at present to do with, and is thus constructed. (*fig.* 2.)

This, it will be observed, contains the inch, the half-inch, the quarter-inch, and the eight of an inch scales, and the application is thus : — Suppose we wish to represent upon paper a line of 24 feet from a scale of the tenth of an inch to a foot. As the inch is divided into ten parts, extend the compasses from the second to the fourth small division on the left upon the *inch scale,* so shall that distance represent the 24 feet as required. If we wish to represent it upon a smaller scale, say the 20th of an inch to a foot, then take from the second to the fourth small division from the *half-inch scale.* Or, if from a still smaller scale, say the 40th of an inch to a foot, take twenty-four divisions from the *quarter-inch scale,* in like manner.

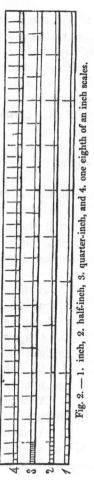

Fig. 2. — 1. inch, 2. half-inch, 3. quarter-inch, and 4. one eighth of an inch scales.

So shall the three lines in the margin represent 24 feet each, taken from different scales. But the same proportion would hold good if, instead of feet, it were inches, yards, or any other measure, as will hereafter be made plain.

Section II.

DEFINITIONS.

1. When a straight line standing on another straight line makes the adjacent angles equal to one another, each of the angles is called a right-angle; and the straight line which stands on the other is called a perpendicular to it. Thus we see that the inclination of one straight line to another forms an *angle*.

2. An obtuse angle is that which is greater than a right-angle.

3. An acute angle is that which is less than a right-angle.

4. A circle is a plane figure contained by one line, called the circumference.

5. A diameter of a circle is a straight line drawn through the centre, and terminated both ways by the circumference.

6. Rectilineal figures are those which are contained by straight lines.

7. Trilateral figures, or triangles, by three straight lines.

8. Quadrilateral, by four straight lines.

9. Polygons, by more than four straight lines.

10. A square is that which has all its sides equal and all its angles right-angles.

11. An oblong is that which has all its angles right-angles, but has not all its sides equal.

12. A rhombus is that which has all its sides equal, but its angles are not right-angles.

13. A rhomboid is that which has opposite sides equal to one another, but all its sides are not equal, nor its angles right-angles.

Section III.

PROBLEMS.

1. To bisect or divide a right line given as A B into two equal parts.

From both ends of the given line (viz. A and B in *fig.* 3.), with any radius greater than half its length, describe two arcs that may cross each other in two points, as at D and E; then join those points D, E with a right line, and it will bisect the line A B in the middle at C; viz. it will make A C equal C B, as was required.

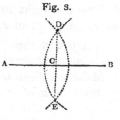

Fig. 3.

2. To draw a right line, as E D in *fig.* 3., parallel to a given right line A B, in *fig.* 4.

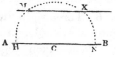

Fig. 4.

Take any convenient point in the given line, as at C; make C X radius, and with it, upon the point C, describe a semicircle, as H M X N; then make the arc H M equal to the arc N X; through the points M and X draw the right line E D, and it will be parallel to the line A B, as was required.

3. To let fall a perpendicular, as C X (*fig.* 5.), upon a given right line A B, from any assigned point that is not in it, as from C.

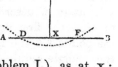

Fig. 5.

Upon the given point C describe such an arc of a circle as will cross the given line A B in two points, as at D and F; then bisect the distance between those two points D, F (by Problem I.), as at X; draw the right line C X, and it will be the perpendicular required.

4. To erect or raise a perpendicular upon the end of any

given right line, as at B in *fig*. 6., or upon any other point
assigned in it.

Upon any point taken at pleasure out of the given line,
as at C, describe such a circle as will

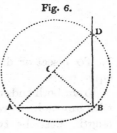

Fig. 6.

pass through the point from whence
the perpendicular must be raised, as
at B (viz. make C B radius); and from
the point where the circle cuts the
given line, as at A, draw the circle's
diameter A C D; then from the point
D draw the right line D B, and it will
be the perpendicular, as was required.

5. Upon a right line given, as A B in *fig*. 7., to describe an equilateral triangle.

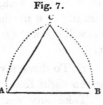

Fig. 7.

Make the given line radius, and with
it, upon each of its extreme points or
ends, as at A and B, describe an arc, viz.
A C and B C; then join the points A, C
and B C with right lines, and they will
make the triangle required.

6 Three right lines being given (*fig*. 8.) to form them
into a triangle, provided any two
of them taken together shall be
longer than the third.

Fig. 8.

Let the given lines be A B, C B, and A C.

Make either of the shorter lines, as A C in *fig*. 9., radius,
and upon either end of the longest
line, as at A, describe an arc; then

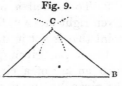

Fig. 9.

make the other line C B radius, and
upon the other end of the longest
side, as at B, describe another arc,
to cross the first arc, as at C; join   A
the points A, C and B, C with right lines, and they will form
the triangle required.

7. Upon a given right line, as A B in *fig*. 10., to form a
square.

Upon one end of the given line, as at B, erect the per-

pendicular B D, equal in length with
the given line ; viz. make B D = A B ;
that being done, make the given line
radius, and upon the points A and D
describe equal arcs to cross each
other, as at C; then join the points
C, A and C, D with right lines, and
they will form the square required.

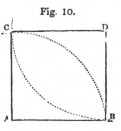

Fig. 10.

8. Two unequal right lines being given (*fig.* 11.) to form
or make of them a right-
angled parallelogram.

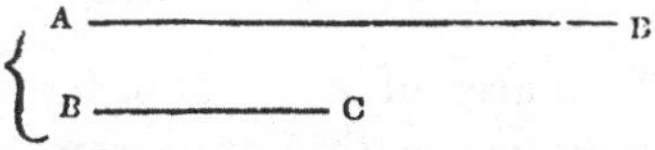

Let the lines be A B and
B C.

Fig. 11.

A ———————— D
B ————— C

Upon one end of the longest line, as at B in *fig.* 12.,
erect a perpendicular of the same
length with the shortest line B C,
in *fig.* 11. ; then from the point
C draw a line parallel to, and of
the same length as, A B ; viz. make D C = A B ; join D A
with a right line. and it will form the oblong or parallelogram required.

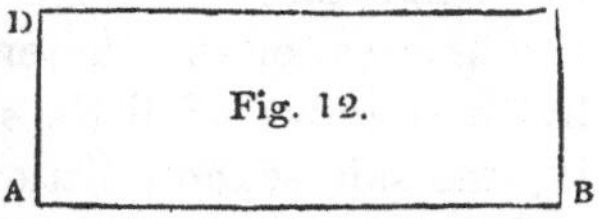

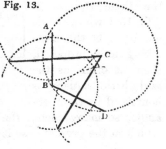

Fig. 12.

9. Three points being given, not in a straight line, to
find the centre of a circle
whose circumference shall
pass through them, as the
points A, B, D, in *fig.* 13.

Join the points A, B and B, D
with right lines ; then bisect
both these lines, and the
point where the bisecting lines
meet, as at C, will be the
centre of the circle required.

Fig. 13.

These nine Problems are thought to be sufficient to
exercise the young practitioner, and bring his hand to the
right management of the ruler and compasses, wherein we
would advise him to be very ready and exact.

CHAP. IV.

MENSURATION.

SECTION I,

MENSURATION OF SUPERFICIES.

THE area of any plane figure is the measure of the space
contained within its extremes or bounds, without any regard
to thickness.

The area, or the content of the plane figure, is estimated
by the number of little squares that may be contained in
it; the side of those little measuring squares being an inch,
a foot, a yard, or any other fixed quantity. And hence
the area or content is said to be so many square inches, or
square feet, or square yards, &c.

Thus, if the figure to be measured be the rectangle
A B C D (*fig.* 14.), and
the little square, E
(*fig.* 15.), whose side is
one inch, be the mea-
suring unit proposed,
then, as often as the
said little square is
contained in the rect-

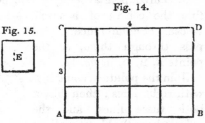

Fig. 14.

Fig. 15.

angle, so many inches the rectangle is said to contain,
which in the present case is 12.

PROBLEM I. To find the Area of any Parallelogram,
    whether it be a Square, a Rectangle, a Rhombus, or
    a Rhomboid.

Rule. *Multiply the length by the perpendicular breadth, or
height, and the product will be the area.*

Ex. 1. To find the area of a parallelogram (*fig.* 16.) whose length is 25 yards, and breadth is 15 yards.

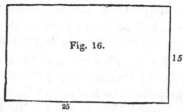

Fig. 16.

15

25

25 length
15 breadth
———
125
25
———
Answer 375 yards.

Project the same upon a scale the 10th of an inch to the yard.

Ex. 2. Find the area of a rectangular board (*fig.* 17.) whose length is 12½ feet, and breadth is 9 inches.

Reduce the inches to the decimal of a foot, and the half foot to a decimal of the same denomination, then proceed as in the last example.

12½ ft. = 12·5
9 in. = ·75
———
625
875
———
9·375 Ans. 9 ft. ·375 dec., or 9⅜ ft.

The figure given is in the proportion of ⅛ of an inch to a foot.

Fig. 17.

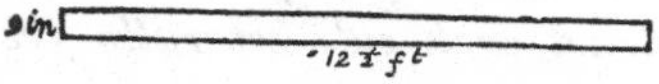

in 12½ ft

Ex. 3. To find the content of a piece of land, in the form of a rhombus (fig. 18.), 120 yards long, and perpendicular height 75 yards.

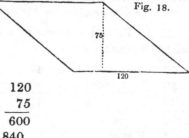

Fig. 18.

75

120

120
75
———
600
840
———
9000 Ans. 1 acre and 4060 yds.

Project the same 100 yards to the inch.

H 3

PROBLEM II.   To find the Area of a Triangle.

Rule 1. *Multiply the base by the perpendicular height,
and half the product will be the area.   Or, multiply the one
of these dimensions by half the other.*

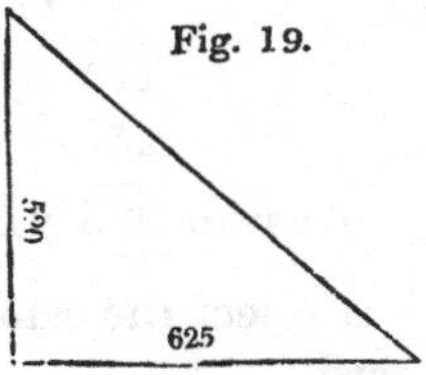

Ex. 1. Find the area of a triangle
(*fig.* 19.) whose base is 625 links, and
perpendicular height 520 links.

Here 625 × 260 = 162500 links, or
1 acre 2 roods 20 perches.

Project the figure to a scale of ¼ of
an inch to 100 links.

Fig. 19.

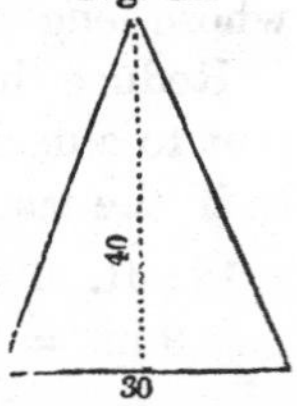

Ex. 2. Find the area of a triangle (*fig.* 20.)
whose base is 30 ft., and perpendicular 40 ft.

Fig. 20.

$$30$$
$$40$$

divide by 2) $\overline{1200}$ = 600 ft. Ans.   Or, to
bring it into yards, divide by 9, the square
feet in a yard, thus, $\frac{600}{9} = 66\frac{2}{3}$ square yards.

Project the same from the scale of 40 divisions to the
inch.

Rule II.   When the three sides are given. *Add all
the three sides together, and take half that sum.   Next, subtract each side severally from the said half sum, obtaining
three remainders.   Lastly, multiply the said half sum and
those three remainders all together, and extract the square
root of the last product for the area of the triangle.*

Ex. 1. To find the area of a triangle whose three sides
are 20, 30, and 40.

20	45	45	45
30	20	30	40
40	25	15	5 three remainders.

2)90

45 half sum.

Then 45 × 25 × 15 × 5 = 84375.

The square root of which is 290·4737, the area.

PROBLEM III.　To find the Area of a Trapezium.

Rule.　*Divide the trapezium into two triangles by a diagonal; then find the area of these triangles, and add them together.*

Note. If two perpendiculars be let fall on the diagonal from the other two opposite angles, and the sum of these perpendiculars be multiplied by the diagonal, half the product will be the area of the trapezium.

Ex. 1. To find the area of the trapezium (*fig.* 21.) whose diagonal is 42, and the two perpendiculars on it 16 and 18.

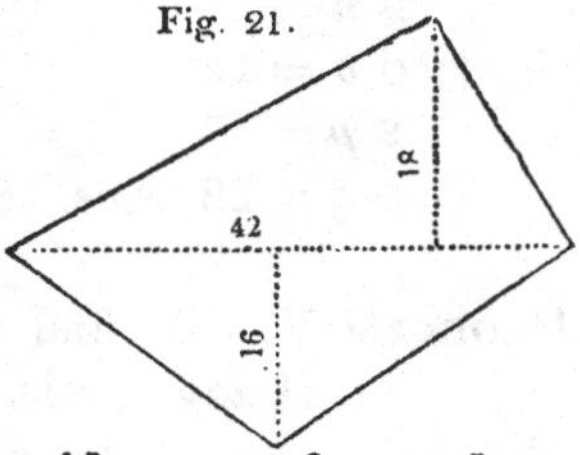

Here 16 + 18 = 34 : the half of which is 17.

Then 42 × 17 = 714, the area.

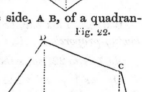

Fig. 21.

Ex. 2. In measuring along one side, A B, of a quadrangular field (*fig.* 22.), that side and the two perpendiculars let fall on it from the two opposite corners, measured as below, required the content.

Fig. 22.

A P = 110
A Q = 745
A B = 1110
C P = 352 　Here 110 × 352　　-　　-　　= 38720
D Q = 595　　745 − 110 = 635 × $\overline{352 + 595}$ = 601345
　　　　　　1110 − 745 = 365 × 595　　 = 217175
　　　　　　　　　　　　　　　　2) 857240
　　　　　　　　　　　　　Ans. 428620

PROBLEM IV.　To find the Area of an irregular Polygon.

Rule.　*Draw diagonals dividing the proposed polygon into trapeziums and triangles. Then find the area of all these separately, and add them together for the content of the whole polygon.*

H 4

Ex. To find the content of the irregular figure A B C D E F G (*fig.* 23.), in which are given the following diagonals and perpendiculars, viz. :

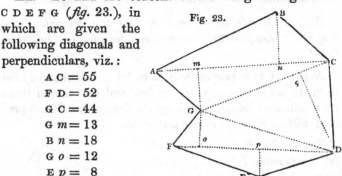

Fig. 23.

    A C = 55
    F D = 52
    G C = 44
    G m = 13
    B n = 18
    G o = 12
    E p = 8
    D q = 28 Ans. 18785

PROBLEM V. To find the Diameter and Circumference of any Circle, the one from the other.

R ule. *This may be done nearly by either of the two following proportions, viz. :*

As 7 is to 22, so is the diameter to the circumference.

Or, as 1 is to 3·1416, so is the diameter to the circumference. (See p. 122.)

Ex. 1. To find the circumference of a circle whose diameter is 20.

By the first rule, as $7 : 22 :: 20 : 62\frac{6}{7}$, the answer.

By the second rule, as $1 : 3\cdot1416 :: 20 : 62\frac{6}{7}$, the same.

Ex. 2. If the circumference of the earth be 25,000 miles, what is its diameter?

As $3\cdot1416 : 1 :: 25000 : 7957\frac{3}{4}$ nearly, the diameter.

PROBLEM VI. To find the Area of a Circle.

Rule I. *Multiply half the circumference by half the diameter. Or, multiply the whole circumference by the whole diameter, and take a quarter of the product.*

Rule II. *Square the diameter, and multiply that square by the decimal ·7854 for the area.*

Ex. 1. To find the area of a circle whose diameter is 10 and circumference 31·416.

By the 1st Rule.

$$\begin{array}{r} 31·416 \\ 10 \\ \hline 4\,)\overline{314·16} \\ \hline 78·54 \quad \text{The area.} \end{array}$$

By Rule 2d.

$$\begin{array}{r} ·7854 \\ 100 = 10^2 \\ \hline 78·54 \end{array}$$

Ex. 2. Find the area of a circle whose diameter is 7 and circumference 22.

$$\begin{array}{r} 22 \\ 7 \\ \hline 4\,)\overline{154} \\ \hline 38·5 \; Ans. \end{array}$$

Ex. 3. How many square yards are in a circle whose diameter is $3\frac{1}{2}$ feet.

Here $(3·5)^2 \times ·7854 \div 9 = 1·069015$ Ans.

Ex. 4. How many square yards of cement are there in a basin for holding water in a garden, whose diameter is 10 feet, and its depth 3 feet.

Here, by Ex. 1., the area of the bottom of the basin is 78·54 feet.

The circumference is $31·416 \times 3 = 94·248$ feet.

Then $(78·54 + 94·248) \div 9 = 19·198$ square yards.

Suppose the price to be 2*s.* 6*d.* per yard, then the cost would be 2·399*l.*, or nearly 2*l.* 8*s.*

The cost of such a basin built of brick, and so covered, may, by this example, be easily ascertained.

PROBLEM VII. To find the Area of an Ellipsis or Oval.

Rule. *Multiply the transverse diameter by the conjugate,*
and multiply the product by
·7854; this last product is the
area of the oval.

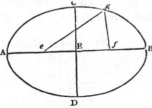

Fig. 24.

Ex. What is the area of the
oval (*fig.* 24.) whose transverse
diameter A B is 40, and the conjugate diameter C D is 20 feet?

Here 40 × 20 × ·7854 = 628·32 feet, the answer.

To construct an Oval mechanically upon the Ground, &c.

Fix two pins equally distant from the centre, as *e f* in
fig. 24., round which pass a double line extending to *g*,
which, carried round, will describe the oval.

Having the Length and Breadth of an Oval given to find
the Distance of each Focus (*e, f*) from the Centre E.

Rule. *From the square of half the transverse subtract the*
square of half the conjugate, and the square root of the dif-
ference will be the distance of each focus from the middle or
common centre of the ellipsis; i. e. $\sqrt{(\overline{\sqrt{A E}}|^2 - \overline{\sqrt{E D}}|^2)} =$
E *e* or E *f*.

Thus let the dimensions stand as above.
    Then 20 = half the transverse diameter,
        and 10 = half the conjugate;
    therefore 400 = the square of 20,
        also   100 = the square of 10,
                    300 the difference, and the square root
extracted will produce 17 feet 5 inches nearly, or the
length of the line E *e* or E *f*.

Section II.

MENSURATION OF SOLIDS.

The measure of a solid is called its solidity, capacity, or content.

1728 cubic inches make one cubic foot $= \overline{12}|^3$.

27 cubic feet make one cubic yard $= 3|^3$.

Problem I. To find the solid Content of any Prism or Cylinder.

Rule. *Find the area of the base or end, and multiply it by the length of the prism or cylinder for the solid content.*

Ex. 1. Find the solid content of a cube whose side is 28 inches.

Thus $28 \times 28 \times 28 = 21952$ in. $= 12 \cdot 7$ cubic feet. Ans.

Ex. 2. How many cubic feet are there in a block of marble, its length being 3 feet 2 inches, its breadth 2 feet 8 inches, and thickness 2 feet 6 inches?

Here reduce the inches to the decimal of a foot, then the work will stand thus: —

$$
\begin{array}{r}
3 \cdot 166 = \text{length.} \\
2 \cdot 666 = \text{breadth.} \\
\hline
18996 \\
18996 \\
18996 \\
6332 \\
\hline
8 \cdot 440556 \\
2 \cdot 5 = \text{depth.} \\
\hline
42202780 \\
16881112 \\
\hline
21 \cdot 1013900 \ \text{solid content.}
\end{array}
$$

Ans. $21 \cdot 10139$ cubic feet.

Ex. 3. How many gallons of water will a cistern contain whose dimensions are the same as in the last example; 282 cubic inches being contained in the gallon.

$$\text{Thus } 38 \text{ length} = 3 \text{ ft. } 2 \text{ in.}$$
$$32 \text{ breath} = 2 \text{ ft. } 8 \text{ in.}$$

$$
\begin{array}{r}
\underline{76} \\
114 \\
\underline{1216} \\
30 \text{ depth} = 2 \text{ ft. } 6 \text{ in.}
\end{array}
$$

$$282)\overline{36480}(129. \quad \text{Ans. } 129\tfrac{17}{47} \text{ gallons.}$$

$$
\begin{array}{r}
282 \\
\hline
\cdot 828 \\
564 \\
\hline
2640 \\
2538 \\
\hline
\cdot 102
\end{array}
$$

$$\frac{\cdot 102}{282} = \tfrac{17}{47}, \text{ or } \frac{21\cdot10139 \times 1728}{282} = 129\tfrac{17}{47}.$$

PROBLEM II. To find the Content of any Pyramid or Cone.

Rule. *Find the area of the base, and multiply the area by the perpendicular height, then take $\frac{1}{3}$ of the product for the content. Every cone is one-third of the circumscribing cylinder of the same base and altitude.*

Ex. 1. Required the solidity of the square pyramid, each side of its base being 30, and its perpendicular height 20.

Thus $30 \times 30 \times 20 = 18000 \div 3 =$ Ans. 6000.

Ex. 2. Find the content of a triangular pyramid, its height being 14 feet 6 inches, and the three sides of its base 5, 6, 7 feet.

Thus the area of the base will be found to be

14·7 nearly,
14·5 the height.
735
588
147
3) 213·15
71·05 solid content.

Ans. 71·05 cubic feet.

PROBLEM III.   To find the Surface of a Sphere.

Rule I. *Multiply the circumference of the sphere by its diameter, and the product will be the whole surface of it.*

Rule II. *Multiply the square of the diameter by 3·1416, and the product will be the surface.*

Ex. 1. Required the convex surface of a sphere whose diameter is 7 and circumference 22.

22 circumference,
7 diameter,
Ans. 154 the surface.

Ex. 2. Required the area of the whole surface of the earth, its diameter being $7957\frac{3}{4}$ miles, and its circumference 25,000 miles.

7957·75
25000
3978875000
1591550
198943750·00

Ans. 198,943,750 square miles.

PROBLEM IV. To find the Solidity of a Sphere or Globe.

Rule I. *Multiply the surface by the diameter, and take* $\frac{1}{6}$ *of the product for the content.*

Rule II. *Multiply the cube of the diameter by the decimal* ·5236 *for the content.*

Ex. 1. To find the content of a sphere whose axis is 12.

Thus 12 × 12 × 12 = 1728, the cube of 12.

$$
\begin{array}{r}
·5236 \\
\hline
10368 \\
5184 \\
3456 \\
8640 \\
\hline
904·7808 \text{ Ans.}
\end{array}
$$

Ex. 2. Find the solid content of the globe of the earth, supposing its circumference to be 25,000 miles.

circum.   diam.

Here 25000 × 7957·75 = Ans. 263,858,149,120 cubic miles.

SECTION III.

PRACTICAL QUESTIONS

RELATING TO GARDENING, FORESTRY, AND FARMING.

Questions in Gardening.

Ex. 1. SUPPOSE a kitchen-garden of a rectangular form (*fig.* 25.) is to be laid out, to contain just an acre; the length being, given (80 yards), what would be the width? — And what would the outside slips contain, the breadth being 7 yards?

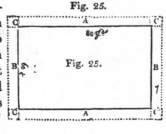

Fig. 25.

The length 80)4840 the square yards in an acre.

yds. 60·5 the width required.

Then 2 ×   80 × 7 = 1120 area of the slips A, A.

2 × 60·5 × 7 = 847 area of the slips B, B.

4 ×   7² = 196 area of the little squares

C, C, C, C.

Square yards 2163 = 1 R. 31$\frac{6.5}{117}$ P, area of the slips.

Ex. 2. Suppose an acre of land was to be planted with trees at 4 feet apart; what number would it take?

Here the practitioner must bear in mind that the ground is squared around each plant, as will appear evident by the diagram in the margin (*fig.* 26.), which shows that 12 trees might be planted at 4 feet apart upon an area of 192 square feet.

Fig. 26.

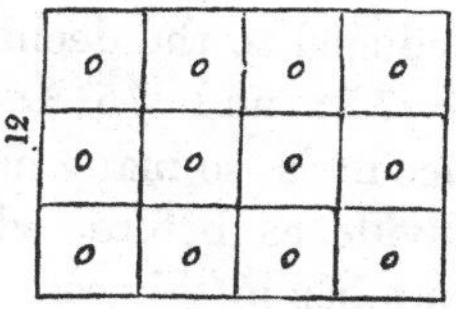

Hence the process is as follows:

Multiply 4840, the square yards in an acre, by 9, the square feet in a yard.

```
              4840
                 9
4² = 16)43560 (2722½
              32
              ---
              115
              112
              ---
            · ·36
              32
              ---
              40
              32
              ---
              ·8 Ans. 2722½, number of trees.
```

Ex. 3. Required the number of trees to plant an acre at two feet six inches from tree to tree.

Here the number of square feet in the acre, as shown

before, is 43560; therefore divide this sum by the square of two feet six inches.

Thus $2{\cdot}5 \times 2{\cdot}5 = 4{\cdot}25$, and $\dfrac{43560}{4{\cdot}25} = 10249$, number of trees, Ans.

The learner will recollect that ${\cdot}5$ is the decimal of $\frac{1}{2}$, and in this case is the decimal of a foot, for it is $6{\cdot}0 \div 12 = 5$. It is hoped he will recollect the reason too why two ciphers are affixed to the dividend; if not, he had better refer back to division of decimals.

Ex. 4. How many coleworts will stand upon an acre at the distance of nine inches from plant to plant?

Here 43560, the square feet as before, and nine inches reduced to the decimal of a foot is ${\cdot}75$, for it is $= 9{\cdot}0 \div 12 = {\cdot}75$; and ${\cdot}75^2 = {\cdot}5625$, now here being four places of decimals, so many must be affixed to the dividend, and then divide as before, which will make the quotient a whole number in this case.

Hence $\dfrac{43560}{{\cdot}5625} = 77{,}440$ plants, Ans.

It is requested that the reader will make himself well acquainted with the nature of the three last examples: their practical usefulness is obvious.

Ex. 5. Suppose a garden wall is to be built one hundred yards long, and ten feet high above the ground, in nine-inch work, and a foundation of two feet six inches deep in fourteen-inch work; what number of bricks will be required to complete the work, and how many rods of work will it contain?

Before showing the calculation it may be necessary to premise that a square yard of nine-inch work contains about one hundred bricks; but though not exactly the case, is near enough for all practical purposes. The bricklayer's work is moreover usually estimated by the rod of $16\frac{1}{2}$ feet square, of a brick and a half (14 inches) thick.

Now from this data we will proceed to the calculation :

Length. Height. Sq. ft.

$300 \times 10 = 3000$ of nine-inch work,

$300 \times 2 \cdot 5 = 750$ of fourteen-inch work, but 1125 reduced to nine-inch work.

Therefore

$3000 + 1125 = 4125 \div 9 = 458 \cdot 33$ square yards,

$$\frac{100 = \text{numb. of bricks in a yd.}}{45833}$$

Also 4125 divided by $272 \cdot 25$, the square feet in a rod, will produce $15 \cdot 151$ square rods.

Hence the answer $\left\{ \begin{array}{l} 45833 \text{ bricks required.} \\ 15 \cdot 151 \text{ rods of work.} \end{array} \right.$

Ex. 6. A sunk fence is 500 feet long, six feet in depth, and eight feet over at top; how many cart-loads (27 feet to the load) have been removed in this cutting?

Here the section, as in the margin (*fig.* 27.), is a right-angled triangle ; therefore $8 \times 6 \div 2 = 24$, the area, which, multiplied by the length, 500, gives 12000, the number of cubic feet.

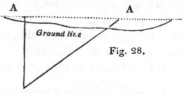

Fig. 27.

Also $\dfrac{12000}{27} = 444$ loads 12 feet, Ans.

Observe. No allowance in the excavation is made here for a brick wall, which would have to be added.

It may not be amiss here to hint to the young gardener having such work to superintend, that by judicious management (in many cases) a great deal of labour in carting, &c. may be spared by depositing the matter dug out upon the adjacent ground.   Thus, suppose the ground was disposed as shown in *fig.* 28., the spaces A A might with propriety be filled up. It should be borne in mind, however, that the desired effect must not be sacrificed to a saving of labour, as the object of a sunk fence is to make the boundary perfectly illusory from particular points of view.

Ex. 7. If a gravel walk is to be made 500 yards in length, seven feet in width, and six inches in depth, how many cubic yards of gravel should be ordered for the purpose?

In this case let the whole dimensions be brought into feet and decimals.

Consequently the length would be   - 1500 feet.
           the breadth   -   -   7 feet.
           the depth of gravel   -   ·5 decimal.

Then $1500 \times 7 \times ·5 = 5250$ cubic feet, which number, divided by 27, the solid feet in a yard, gives the

Ans. 194 yards and 12 feet.

To measure Timber.

The young practitioner ought now to be competent to find the superficial or solid content of any board or squared solid piece of timber by what has already been taught. But the *usual* way to measure *round timber* trees is to girth them about the middle with a string, and take the fourth part of that girth for the side of a square, by which they measure the piece of timber as if it was square.

But that this method is erroneous, the learner himself could easily prove. The universal practice, however, is to measure all rough round timber in the manner stated.

If the tree be very irregular in thickness, it is usual to girth it in several places, and take the mean of those girths, which is considering the tree as a cylinder.

Ex. 8. If a piece of timber be 96 inches in circumference (or girth) in the middle, and 18 feet long, how many feet are contained therein?

A fourth part of 96 is 24: and $24^2 = 576$ area of section.

Hence $\dfrac{576 \times 18}{144} = 72$ feet solid, Ans.

Ex. 9. What is the content of a timber tree whose length is 20 feet, and girth at the root end 92 inches, and the other end 60 inches?

Here $\dfrac{92+60}{2}=76$, mean girth; and $\dfrac{76}{4}=19$ side of the square; then, $\dfrac{19\times19\times20}{144}=50\cdot1388$ cubic feet, Ans.

N.B. The learner will understand that in the two last examples the reason of dividing by 144 is, because the area in square inches is multiplied by the length in feet; but if the said area had been brought into feet and decimals the product alone would have been the answer. On the other hand, if the length had been brought into inches the division should have been by 1728.

Ex. 10. A tree shaped as in *fig.* 29. girths in four different places, respectively 100, 90, 60, and 32 inches, and the length 35 feet: how many cubic feet does it contain?

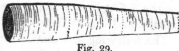

Fig. 29.

Here $\dfrac{100+90+60+32}{4}=\dfrac{282}{4}=70\cdot5$ mean girth; and $\dfrac{70\cdot5}{4}=17\cdot625=$ side of the square; also, $\dfrac{\overline{17\cdot625}\,|^2\times35}{144}=$ $75\frac{1}{2}$ cubic feet.

Ex. 11. Suppose an oak tree is of the following growth and actual dimensions, and that the forester is required to find the solid contents of the stick, as receivable,

1st, into merchants' dockyards:

2dly, into Her Majesty's dockyards:

3dly, when sold by the quarter girth; in order to exhibit in his books the actual dimensions of such a piece of "goods" having "stops," and in regard also to the relative proportions between the three practised modes of mensuration. (See *Chatfield's Measuring Companion*, p. 10—15.)

Fig. 29a.

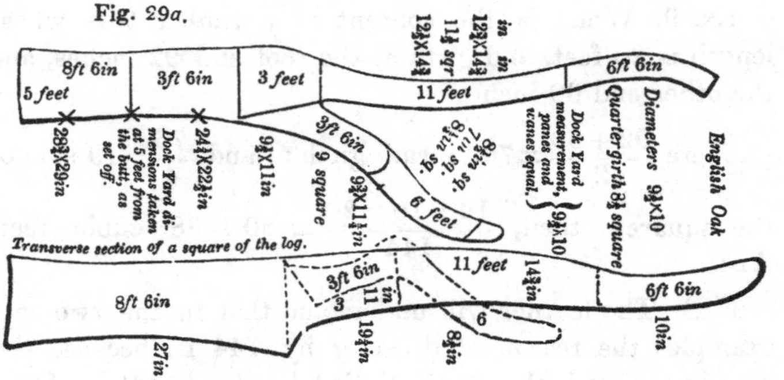

The under figure represents a transverse section of a square of the timber. The figures in the diagrams should agree with the data in the calculation.

The results of these measurements prove the correctness of the areas referred to in the diagrams, exclusive of the butt lengths, 5 feet, and 3 feet 6 inches, which are taken; *i. e.* the 5 feet by a systematic rule, which we will presently explain, and the $3\frac{1}{2}$ feet consequent thereon.

The following calculation presents a full exposition of the diagrams, with the dimensions taken in conformity to the several systems as established for measuring.

No. 1.			No. 3.			No. 2.		
Length, feet.	Squares, inches.	Contents, feet.	Length, feet.	Quarter girth in inches.	Contents, feet.	Length, feet.	Squares, inches.	Contents, feet.
$6\frac{1}{2}$ ×	$9\frac{1}{2} \times 10$	= 4	$6\frac{1}{2}$ ×	$8\frac{1}{2}$	= 3	$6\frac{1}{2}$ ×	9×10	= 4
11 ×	$12\frac{3}{4} \times 14\frac{3}{4}$	= 14	11 ×	$11\frac{1}{2}$	= 10	11 ×	$12\frac{1}{2} \times 14\frac{3}{4}$	= $13\frac{1}{2}$
3 ×	$22 \times 19\frac{1}{4}$	= $8\frac{1}{2}$	3 ×	$19\frac{1}{2}$	= $5\frac{1}{2}$	3 ×	$18\frac{1}{4} \times 20$	= $7\frac{1}{2}$
6 ×	$8\frac{1}{2} \times 8\frac{1}{2}$	= 3	6 ×	7	= 2	6 ×	$8\frac{1}{4} \times 8\frac{1}{4}$	= $2\frac{1}{2}$
$3\frac{1}{2}$ ×	$9\frac{3}{4} \times 11\frac{1}{4}$	= $2\frac{1}{2}$	$3\frac{1}{2}$ ×	9	= $1\frac{1}{2}$	$3\frac{1}{2}$ ×	$9\frac{1}{2} \times 11$	= $2\frac{1}{2}$
$8\frac{1}{2}$ ×	26×27	= 41	$8\frac{1}{2}$ ×	$21\frac{1}{2}$	= 27	$3\frac{1}{2}$ ×	$24\frac{1}{4} \times 22\frac{1}{2}$	= 13
						5 ×	$28\frac{1}{4} \times 29$	= 29
$38\frac{1}{2}$	Total	73	$38\frac{1}{2}$	Total	49	$38\frac{1}{2}$	Total	72

Hence we see that

No. 1. as receivable in merchants' dockyards, and by the timber trade, contains - - - 73 feet.
No. 3. ditto, by girth or string measure - 49 —
No. 2. ditto, into Her Majesty's dockyards - 72 —

Whence it is obvious, Nos. 1. and 3. coincide as nearly as possible for actual operations to the diagrams of the areas. And No. 2. also, deducting in the last two lengths, $3\frac{1}{2}$ and 5, with No. 1., and comparing the remainder of the contents with those of Nos. 1. and 3., there is a near approximation of the several measures. To account for the rapid increase in the proportion of contents in No. 2., it will be seen that a 5 feet length is taken from the butt, which occasions an approximation of total contents to No. 1.; the measurement of 5 feet from the butt has long been an established rule in the Royal dockyards; and it results, that Nos. 1. and 2., upon the whole, do not vary in amount. This will be the case in any quantity of oak sticks tolerably well hewn, as described in the proportions of the diagrams of the areas. In these calculations and remarks, a system is comprised for measuring round timber, the dimensions of which are resolvable by the quarter girth, as treated in the other examples of this work.

If the butts of trees be shaky or foxy, determine the length of the injury likely to be sustained from the butt, and cube that length by the squares at the first measuring place. Thus will the allowance for defects bear an equitable proportion to the contents. All timber requiring more than one third deduction is considered unmarketable.

In general, to measure trees,

Multiply the feet in the length by the square of the inches, and divide for cubic feet by 2304. *Or when an 8th, or 10th, for bark is allowed, by* 3009 *or* 2845.

A standing tree is measured by *squaring $\frac{1}{4}$th of the girth, and multiplying it by the height of the trunk*, as in Example 10.

The length of a tree into the square of its mean girth by 1807, is its cubic feet; or for $\frac{1}{8}$th the bark, by 2360 will give its cubic feet.

This log was selected from a large pile of timber, on account of the number of *stops;* for, by having many dimensions, it was the more likely to prove, upon an average, how far the general conclusion might be determined, in regard to the relative proportions between the three practised modes.

To estimate the contents of Ricks.

Ex. 12. Suppose a truss of hay weighing 56 lb., a yard square, and 9 inches in thickness, be cut from the middle of a rick (*fig.* 30.); how many loads, of 36 trusses to the load, will the rick contain, the dimensions

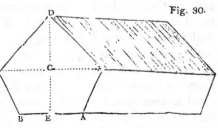

Fig. 30.

being A B = 16 feet, a b = 20 feet, C E = 8 feet, C D = 12 feet, the length = 30 feet.

Now, the area of the section (or end) multiplied by the mean length solves this problem. Therefore the work will stand as follows : —

$$20 + 16 = 36 \times \ 8 = 288$$
$$20 \times 12 = 240$$

528 double the area of the section.

264 the area of the end or section.

30 the length.

7920 the cube content in feet.

Area of the truss = 9 feet, the thickness in decimals ·75 feet; therefore $75 \times 9 = 6·75$, cubic content of the truss.

Here $\dfrac{7920}{6·75} = 1173·33$ trusses = 32 loads and $21\frac{1}{4}$ trusses, Ans.

In calculations of this kind great exactness is not attainable (for reasons unnecessary to be explained); but a useful approximation may always be arrived at, and much nearer the truth than the most practised eye could come to by mere guess.

Hay-ricks are not unfrequently built upon a circular base; in which case they partake of the form of what is called the " parabolic conoid." And to find the solid content, the *rule* is, multiply the area of the base by half the height.

Ex. 13. To find the cubic content of a rick of the form shown in *fig.* 31.

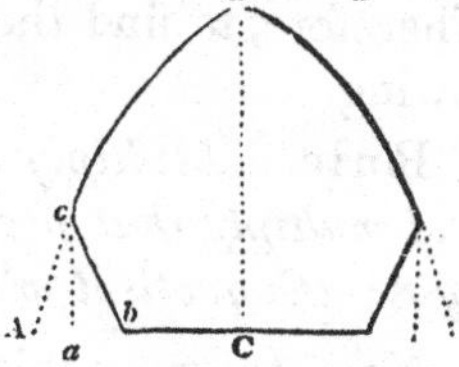

In this case first calculate for the parabolic conoid, and after deduct for the deficiency at the bottom.

Fig. 31.

Let A B $= 10 =$ diam. of the base.

 C D $= 16 =$ the height.

 A $b =$ 2

 $c\,a =$ 4

Then $10 \times 10 \times \cdot 7854 = 78 \cdot 54$ the area of the base.

 and $16 \div 2 =$ 8

      $\overline{628 \cdot 32}$ cubic content of the parabolic conoid.

Also, $\frac{1}{2}$ of $2 \times 4 = 4$, section of the skirting, and $10 \times 3 \cdot 1416 ^* = 31 \cdot 416$, length of ditto.

Again, $31 \cdot 416 \times 4 = 125 \cdot 66$.

   628·32
   125·66
   $\overline{502 \cdot 66}$ the cubic content, as required.

Ricks of corn may be estimated in the same manner, but not so satisfactorily; as both the yield of the grain and the density of the mass of sheaves would have to be come at before an estimate, by measurement, could be made.

To estimate Dung-heaps.

Ex. 14. A heap of dung upon a rectangular base, and form of outline as in *fig.* 32.

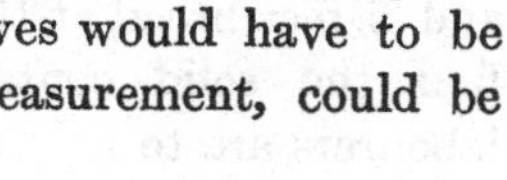

It is required to know how many cart-loads it contains, of two cubic yards to the load, the dimensions being as follow:

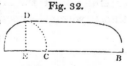

Fig. 32.

---

 * The number 3·1416 is the circumference of a circle whose diameter is 1, as shown in Problem V. p. 104., and note, p. 122.

B C = 15 yards, A C = 10 yards, D E = 8 yards.

Here the section A D C, in *fig.* 32., is in the form of a *parabola;* a figure which is made by the section of a cone, and is two-thirds of its circumscribing parallelogram. Therefore, to find the area of this section, we have the following

Rule. *Multiply the base by the perpendicular height, and multiply that product by 2, and divide the last product by 3 : the quotient will be the area of the parabola.*

Now, to apply this to the question, we have

$$\frac{(A C \times E D)2}{3} = \text{area};$$ for A C = 10 × E D = 8 = 80 × 2 = 160

÷3 = 53·33, area of the section; then, 53·33 × 15, the length, will give 800 cubic yards nearly, and consequently the

Ans. will be 400 loads.

To estimate Road Materials.

Materials prepared for roads are piled up like truncated pyramids, and slope from the ground to the summit at an angle of about 35 degrees, which, for loose materials, is about the angle of repose.

Hence, if a heap of broken granite be 24 feet in length, and 9 feet broad at the base, and its depth be 33 inches, find the solid content in cubic yards, by which the labourers are to be paid for their work ; the crown of the heap being such as a sloping side, at an angle of 35 degrees, will give, from such a pyramidal figure as with the compasses may be constructed from the foregoing data.

If the reader should have any difficulty here, let him carefully read the next chapter; for here, as in other things, we must sometimes anticipate steps of our progress.

CHAP. V.

PRACTICAL TRIGONOMETRY.

SECTION I.

PLANE TRIGONOMETRY.

DEFINITIONS.

1. PLANE TRIGONOMETRY treats of the relations and calculations of the sides and angles of plane triangles.

2. The circumference of every circle is supposed to be divided into 360 equal parts, called degrees, also each degree into 60 minutes, each minute into 60 seconds, and so on.

Hence a semicircle contains 180 degrees, and a quadrant 90 degrees.

3. The measure of any angle is an arc of any circle contained between the two lines which form that angle, the angular point being the centre; and it is estimated by the number of degrees contained in that arc.

Hence a right-angle, being measured by a quadrant, or quarter of a circle, is an angle of 90 degrees; and the sum of the three angles of every triangle, or two right-angles, is equal to 180 degrees. Therefore, in a right-angled triangle, taking one of the acute angles from 90 degrees, leaves the other acute angle; and the sum of two angles in any triangle, taken from 180 degrees, leaves the third angle; or one angle being taken from 180 degrees leaves the sum of the other two angles.

4. Degrees are marked at the top of the figure with a small °, minutes with ′, seconds with ″, and so on. Thus, 57° 30′ 12″ denotes 57 degrees 30 minutes and 12 seconds.

Note. — The sides of every triangle are *proportional* to the sines of the opposite angles. The tangent is a *fourth* proportional to the co-sine, sine, and radius. The secant is a *third* proportional to the co-sine and radius. The co-tangent is a *fourth* proportional to the sine, co-sine, and radius. The co-secant is a *third* proportional to the sine and radius. It is a property of the triangle, that if each of any two of its sides be × by the co-sine of the angle which the side makes with the third side, the sums of the products = that of the third side.

The intention here is to treat of Trigonometry geometrically chiefly; which is accomplished by a pair of compasses, and a scale, with chords, &c., marked thereon.

Fig. 33. will show the reader what is meant by chords, tangents, secants, &c.

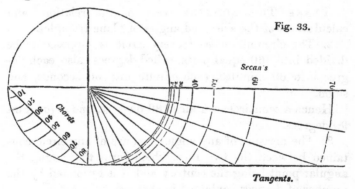

Fig. 33.

Note. — Suppose the diameter of this circle to be 1, its circumference is 3·1416; then, calling D the diameter, C the circumference, A the area, and P = 3·1416, we have

$$\text{D} = \frac{\text{C}}{\text{P}}, \text{ or } \frac{4\text{A}}{\text{C}}, \text{ or } 2\sqrt{\frac{\text{A}}{\text{P}}}; \text{ also } \text{C} = \text{PD}, \text{ or } \frac{4\text{A}}{\text{D}}, \text{ or } 2\sqrt{\text{PA}};$$

$$\text{And } \text{A} = \frac{\text{PD}^2}{4}, \text{ or } \frac{\text{C}^2}{4\text{P}}, \text{ or } \frac{\text{DC}}{4}; \text{ also } \frac{\text{C}}{\text{D}}, \text{ or } \frac{4\text{A}}{\text{D}}, \text{ or } \frac{\text{C}^2}{4\text{A}}.$$

The length of an arc is 8 times the chord of half the arc less the chord of the whole, and divided by 3.

The periphery of an ellipse is $\sqrt{}$ of $\frac{1}{2}$ the sum of the squares of the 2 axes by 3·1416.

The arc of a quadrant 1·578 is 10-9ths the chord 1·4142.

The chord of $\frac{1}{6}$ of a circle whose diameter is 1 is 1·732; of $\frac{1}{4}$, 1·4142; of $\frac{1}{5}$ is 1·1755; of $\frac{1}{6}$ is 1; of $\frac{1}{10}$, 0·68404; of $\frac{1}{12}$, 0·5176.

The length of any circular arc is the radius × by the degrees in the arc × by 0·017453.

The decimal of a degree or 360th of the circle is 0·008726.

The square of the circle is 9·8696; and the square root of the circle is 1·77245.

Ex. 1. In the plane triangle A B C, in *fig.* 34.,

Given $\left\{\begin{array}{l} \text{A B} \qquad 345 \text{ yds.} \\ \text{A C} \quad 174\cdot07 \text{ yds.} \\ \angle\text{A} \quad 37^\circ\ 20' \end{array}\right.$

Required the other side, and the remaining angles.

Fig. 34.

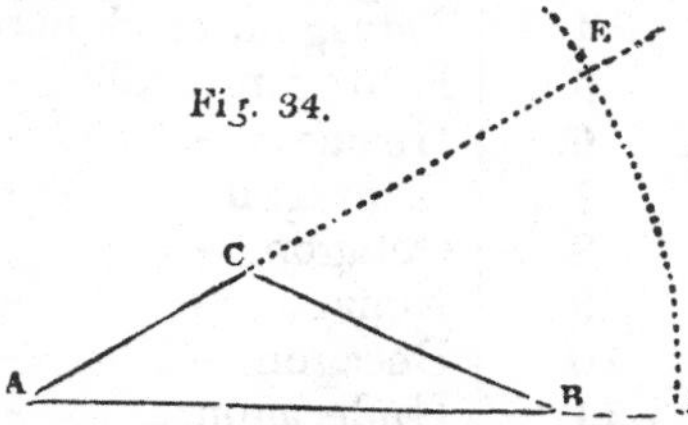

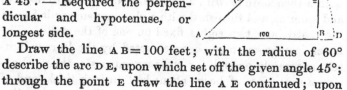

Draw A B = 345 from a scale of equal parts; make the angle A = 37° 20′, which is thus performed : — Take from the scale of chords in your compasses the extent of 60° as radius, then with one foot of your compasses on A describe the arc D E, upon which set off the given angle D E, and through the point E draw the line A E continued; set off A C, and join C B, and it is done.

Then the other parts being measured, they are found to be nearly as follows; viz. the side B C = 232 yards, the angle B = 27°, and the angle C = 115° 30′.

Fig. 35.

Ex. 2. In the right-angled triangle A B C, in *fig.* 35., given the base A B 100 feet, the angle A 45°. — Required the perpendicular and hypotenuse, or longest side.

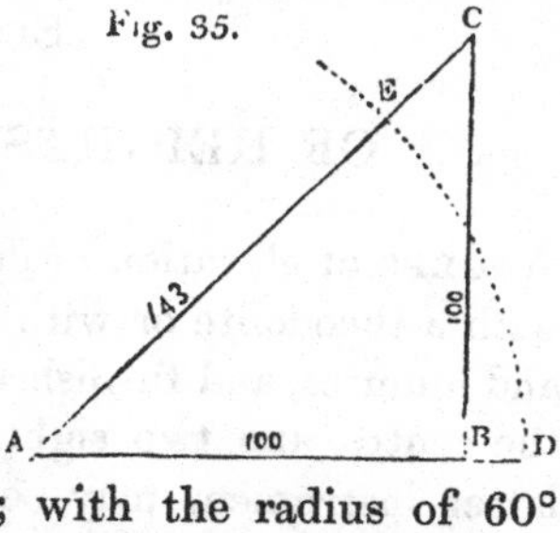

Draw the line A B = 100 feet; with the radius of 60° describe the arc D E, upon which set off the given angle 45°; through the point E draw the line A E continued; upon the point B erect the perpendicular, which will meet the line A E at C. Then the lines A C and B C, measured on the scale of equal parts, will be found to be 143 and 100 feet.

Table of the Areas of Polygons whose sides are 1.

Sides.	Polygons.	Areas.
3	Trigon - - -	0·4330127
4	Tetragon, or Square -	1·0000000
5	Pentagon - -	1·7204774
6	Hexagon - - -	2·5980762
7	Heptagon - -	3·6339124
8	Octagon - - -	4·8284271
9	Nonagon - - -	6·1818242
10	Decagon - - -	7·6942088
11	Undecagon - -	9·3656389
12	Dodecagon - -	11·1961524

All these figures can be inscribed within, and circumscribed about, a circle.

———————

SECTION II.

OF HEIGHTS AND DISTANCES.

ANGLES of elevation or depression are usually taken either with a theodolite or with a quadrant, divided into degrees and minutes, and furnished with a plummet suspended from the centre, and two sights fixed on one of the radii. The latter instrument may be constructed by the learner himself, accurately enough for ordinary purposes, thus. Procure a nice piece of board about half an inch thick, and of a size sufficient to cut out an exact quadrant of a radius of six or nine inches; then divide the arc into 90 divisions, which will be degrees, which subdivide for half-degrees, &c.   Suspend a plummet by a thread from the angular point, and fix two sights on one of the sides, which may be made of any thin piece of metal with small holes perforated.   For taking a steady observation the instrument may be attached to a pointed staff that may be fixed in the ground, but moveable to any elevation or depression.

To take an Angle of Altitude and Depression with the Quadrant.

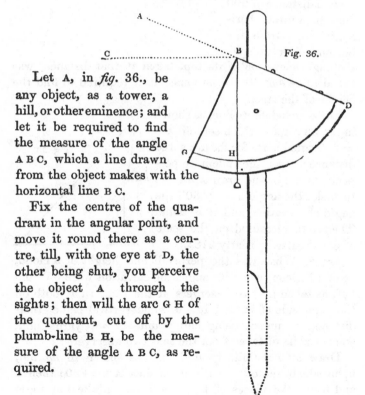

Fig. 36.

Let A, in *fig.* 36., be any object, as a tower, a hill, or other eminence; and let it be required to find the measure of the angle A B C, which a line drawn from the object makes with the horizontal line B C.

Fix the centre of the quadrant in the angular point, and move it round there as a centre, till, with one eye at D, the other being shut, you perceive the object A through the sights; then will the arc G H of the quadrant, cut off by the plumb-line B H, be the measure of the angle A B C, as required.

The angle A B C of depression of any object, A in *fig.* 37., is taken in the same manner, except that here the eye is applied to the centre, and the measure of the angle is the arc G H, on the other side of the plumb-line.

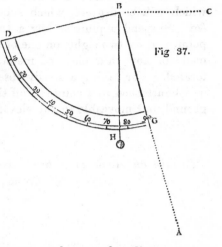

Fig 37.

Ex. 1. Having measured a distance of 200 feet, in a direct horizontal line, from the bottom of a steeple, the angle elevation of its top, taken at that distance, was found to be 47° 30′: from hence it is required to find the height of the steeple.

Draw an indefinite line, as shown in *fig.* 38., upon which set off A C = 200 equal parts for the measured distance; erect the indefinite perpendicular C B, and draw A B so as to make the angle A=47° 30′, the angle of elevation, and it is done. Then C B, measured on the scale of equal parts, is nearly 218¼.

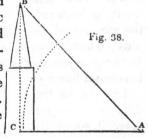

Fig. 38.

Ex. 2. What was the perpendicular height of a cloud, or of a balloon, when its angles of elevation were 35° and 64°, as taken by two observers at the same time, both on the same side of it, and in the same vertical plane, their distance, as under, being half a mile, or 880 yards; and what was its distance from the said two observers?

Draw an indefinite ground line, as shown in *fig.* 39., upon which set off the given distance A B=880; then A and B are the places of the observers. Make the angle

A = 35°, and the angle B = 64°; and the intersection of the lines at C will be the place of the balloon; from whence the perpendicular C D, being let fall,

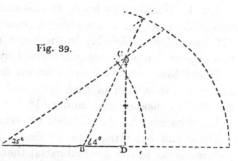

Fig. 39.

will be its perpendicular height. Then, by measurement, are found the distances and height nearly as follows, viz. A C 1631, B C 1041, D C 936.

Ex. 3. Having to find the height of a tree standing on the top of a declivity, I first measured from its bottom a distance of 40 feet, and there found the angle, formed by the oblique plane, and a line imagined to go to the top of the tree, 41°; but after measuring on in the same direction 60 feet further the angle at that point was only 23° 45′. What then was the height of the tree?

Draw an indefinite line for the sloping plane or declivity, as shown in *fig.* 40., in which assume any point A for the bottom of the tree, from whence set off the distance A C

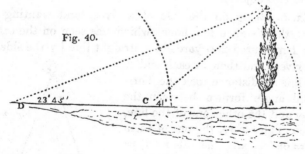

Fig. 40.

= 40, and again C D = 60 equal parts; then make the angle C = 41°, and the angle D = 23° 45′; and the point B where the two lines meet will be the top of the tree. Therefore A B, joined, will be its height, on the same scale.

Ex. 4. Wanting to know the distance between two in-accessible trees, or other objects, from the top of a tower, 120 feet high, which lay in the same right line with the two objects, I took the angles formed by the perpendicular wall and lines conceived to be drawn from the top of the tower to the bottom of each tree, and found them to be 33° for the nearer tree, and 64½° for the more remote. What then may be the distance between the two objects?

Draw the indefinite ground line B D, in *fig.* 41., and perpendicular to it B A = 120 parts; then draw the two lines

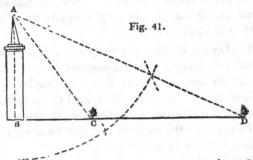

Fig. 41.

Ans. 160 feet.

A C, A D, making the two angles B A C, B A D equal to the given angles 33° and 64½°. So shall C and D be the places of the two objects.

Ex. 5. Being on the side of a river, and wanting to know the distance to a house which was seen on the other side, I measured 200 yards in a straight line by the side of the river, and then at each end of this line of distance took the hori-zontal angle formed between the house and the other end of the line, which angles were, the one of them (A) = 68° 2′, and the other (B) = 73° 15′. What then were the dis-tances from each station to the house?

Draw the line A B, in *fig.* 42., = 200 equal parts. Then draw A C so as to make the angle A = 68° 2′.

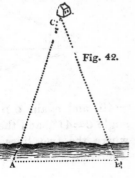

Fig. 42.

and B C to make the angle B = 73° 15′.   So shall the
point C be the place of the house required.

Ans. $\left\{\begin{array}{l} A\ C = 310 \\ B\ C = 290 \end{array}\right\}$ yards.

If we trace on the ground a line A D, in *fig.* 43., as a
continuation of C A, in
fig. 42., and a line B E
as a continuation of C B;
then it is possible to trace
lines B G and A F re-
spectively parallel to A D
and B E, and the point H

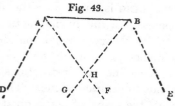

Fig. 43.

where these parallels intersect will indicate the position
of the house C, in *fig.* 42., as if situate on this side of
the stream: that is to say, A H = 310 yards; and B H =
290 yards.

N. B. Horizontal angles cannot of course be taken
with the quadrant.   Therefore, the above example would
require some other instrument for this purpose, as the
theodolite, &c. &c.

Horizontal angles or bearings may be taken in this
manner.   Let B and C, in *fig.*
44., be two objects or two
pickets set up perpendicularly,
and let it be required to take
their bearings, or the angles
formed between them at any
station, A.

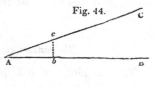

Fig. 44.

Measure one chain length (or any other distance) along
both directions, as to *b* and *c*.   Then measure the distance
b, *c*, and it is done.   This is easily transferred to paper,
by making a triangle A, *b*, *c*, with those three lengths, and
then measuring the angle A.

Section III.

DIALLING.

Dialling is founded on the first motion of the heavenly bodies — or the diurnal motion of the earth on its axis — and its theory depends on the elements of spherical trigonometry. But in an elementary work like this, we shall confine our instructions to what is practicable by a manual operation alone. The plane of every dial represents the plane of some great circle on the earth, and the gnomon of the earth's axis; the vertex of a right gnomon, the centre of the earth or visible heavens; and the plane of the dial is just as far from this centre as from the vertex of this style. The earth itself, compared with its distance from the sun, is considered as a point; and, therefore, if a small sphere of glass be placed upon any part of the earth's surface, so that its axis be parallel to the axis of the earth, and the sphere have such lines upon it, and such planes within it as are required, it will show the hours of the day as truly as if it were placed at the centre of the earth, and the shell of the earth were as transparent as glass.

To construct an Equatorial Dial.

Take a piece of oak plank a foot square, and 1 to 2 inches thick, with two sunk cross bars on the back to prevent warping in the sun. Divide it into a circle of 24 equal parts, beginning at 1 and going on to 12, and again repeated to complete the circle. The two lines of 6 and 12 must be at right angles; the board painted white and the lines black. Set a wire pin in the centre of the circle of the same thickness as the black lines, quite perpendicular, and at right angles with the lines 6 and 12. Find the latitude of the place, say 53°; subtract from 90°=37° or co-latitude; make two boards to this angle; place the dial level on a post, and the face northwards, elevated by the angled board behind, and the hours will be pointed correctly, by the shadow of the wire on the lines, for six months, when the sun is north of the line. In places south of the line, the face of the dial will be reversed, and stand to the south.

Table of the Angles which the Hour Lines form with the Meridian on a Horizontal Dial for every Half Degree of Latitude from 50° to 59° 30′.

Latitude.	A. M. I. XI.		A. M. II. X.		A. M. III. IX.		A. M. IV. VIII.		A. M. V. VII.		A. M. VI. VI.	
° ′	°	′	°	′	°	′	°	′	°	′	°	′
50 00	11	38	23	51	37	27	53	0	70	43	90	00
50 30	11	41	24	1	37	40	53	11	70	51	90	00
51 00	11	46	24	10	37	51	53	24	70	58	90	00
51 30	11	51	24	19	38	4	53	36	71	6	90	00
52 00	11	55	24	27	38	14	53	46	71	13	90	00
52 30	12	00	24	36	38	25	53	58	71	20	90	00
53 00	12	5	24	45	38	37	54	8	71	27	90	00
53 30	12	9	24	54	38	48	54	19	71	34	90	00
54 00	12	14	25	2	38	58	54	29	71	40	90	00
54 30	12	18	25	10	39	8	54	39	71	47	90	00
55 00	12	23	25	19	39	19	54	49	71	53	90	00
55 30	12	28	25	27	39	29	54	59	71	59	90	00
56 00	12	32	25	35	39	40	55	8	72	5	90	00
56 30	12	36	25	45	39	50	55	18	72	12	90	00
57 00	12	40	25	51	39	59	55	27	72	17	90	00
57 30	12	44	25	58	40	9	55	37	72	22	90	00
58 00	12	48	26	5	40	18	55	45	72	27	90	00
58 30	12	52	26	13	40	27	55	54	72	33	90	00
59 00	12	56	26	20	40	36	56	2	72	39	90	00
59 30	13	00	26	27	40	45	56	10	72	44	90	00

In this table the angles formed by the lines for the hour of v. in the morning and that of vii. in the evening, — iv. in the morning and viii. in the evening, &c. are not marked because they are the same as those for vii. in the morning and v. in the evening,—viii. in the morning and iv. in the evening, only they lie on opposite sides of the vi. o'clock hour lines. For the edge of the plane, by which the time of day is marked, is the style of the dial, and it is parallel to the axis of the earth. Hence the line on which this

plane is erected is the substyle ; and the angle included
between the substyle and style is the elevation or height of
the style or gnomon ; moreover this angle is always equal
to the elevation of the pole, or latitude of the place.

From this explanation it is clear that an erect south
dial in 51½° north latitude would be a horizontal dial on
the same meridian 90° southward, which falls in with 38½°
south latitude.

The use of the foregoing table may be easily comprehended : thus, if the place for which a horizontal dial is to
be made corresponds with any latitude in the table, the
angles which the hour lines make with the meridian are
found by inspection.   For example, the hour lines of XI.
and I. must, in the latitude of 56°, make angles of 12° 32′
with the meridian.

If the latitude be not contained in the table, proportional parts may be taken without any sensible error.
Thus, if the latitude be 54° 15′, which would agree with
Milton in Westmoreland, or Kirby Wiske, and the angles
made by the hour lines of XI. or I. be required, as it appears
from the table that the increase of 30′ in the latitude,
namely, from 54° to 54° 30′, corresponds to an increase of
4′ in the hour angle at the centre of the dial, we may
infer, that an increase of 15′ will require an increase of
2′ nearly, and therefore that the angle required will be
12° 16′.

If these instructions be followed, we think every young
gardener may be able to make a dial for the division of his
own time, and the government of his duties at stated
periods of the day.   If by a good chronometer he could
mark the shade of a tall tree at *noon*, by tracing a circle of
some radius within the clear shadow of the stem, he might,
on that circle, set off the angles corresponding to the hour
lines in the foregoing table, and, by planting a flower or a
shrub in the intersection of the hour line with the circle,
exhibit the *dial of paradise.*

CHAP. VI.

MECHANICS.

MECHANICS, in the enlarged sense of the word, includes the whole range of natural philosophy or physics, and developes innumerable principles of the pure and mixed mathematics. But in a limited sense we confine mechanics to those general principles that govern forces as combined with matter. Hence, the sciences of statics and dynamics; upon which, however, this work cannot enter further than to explain a very small number of the principles, that may be serviceable to those for whom we write.

Force either produces or destroys motion. The unit of force may be taken at 1 lb. troy, equal to 22·815 cubic inches of distilled water *, which, divided into 5760 equal parts, the weight of each is a grain troy, and 7000 such grains make 1 lb. avoirdupois. Hence 15 lb. avoirdupois represent a force of 15 units.

When a body is held at rest by two forces, these are said to be equal to one another. Here the forces act in opposite directions, and in the same straight line. We may multiply the forces, but that which counteracts its antagonists exercises double, treble, &c., their intensity if it preserve the equilibrium.

Lines may represent forces in *magnitude*, and also in *direction*. When a third force is required to constitute an equilibrium, if lines be measured from this point in the direction of the forces, so as to contain each a given unit of length as many times as there are units in each force, then these lines will form the

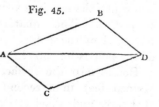

Fig. 45.

* This standard is fixed by act of parliament dated June 24th, 1824: temperature 62° Fahrenheit; barometer to stand at 30 inches.

K 3

adjacent sides and diagonal of a parallelogram. Thus, if
B A, C A, in *fig.* 45. be two forces acting upon the point
A, we determine the magnitude and direction of the force
which will hold them at rest by completing the parallelogram A B D C, and drawing the diagonal A D, which represents both the magnitude and direction of the force
that will keep the others in equilibrium.

Thus we see that *equilibrium* results from the simul-
taneous action of several forces on a body, or a material
point, when they reciprocally destroy each other's action,
and the body, though free to move, remains at rest. Similar
pressure, as noticed above, exhibits force counteracting
force, and entire absence of motion. We thence infer that
all bodies present themselves to us in a state of rest or of
motion. But this last is the change of rectilinear distance
between two points. And as the force producing motion
may act upwards or downwards, &c., we trace motion in a
straight line, bent or curvilinear, accelerated or retarded.
Hence *inertia* is a contingent condition of rest or motion,
yet the earth, in her orbicular and annual motions, and all
the rest of the planets, wheel round the sun without effort
and in ceaseless motion; so that upon her surface nothing
is absolutely at rest. The listless savage at the equator is
moving, when asleep, with a velocity of 1000 feet per second
of time, or 60,000 feet an hour: and the trees there, suspended by their very roots, moving with this velocity, have
not a leaf disturbed. Let but a hurricane sweep the same
region with a force of 120 miles an hour, and all is desola-
tion! The two revolutions of the earth, on its axis daily,
and round the sun in a year, are employed as the standards
of *motion.* The equatorial diameter is seventeen miles
longer than the polar, by which means a mass of 1000 lb.
at the pole is but 995 lb. at the equator.

Bodies near the surface of the earth fall through about
sixteen feet in a second of time. Hence we discover
that force and velocity, time and space, are quantities in a
continual flux, passing through a series of proximate stages.
Thus it is, that the velocity of falling bodies accumulates,

and in the next second it measures 32 feet, which being doubled in the third second gives 64.*

In ascending perpendicularly, the force of gravity, or weight, produces this effect: that a body shot upwards with double velocity rises four times as far as if shot with a single velocity: if with triple velocity, it rises nine times as far.

The *centre of gravity* of a body, or system of bodies, is that point in the body or system, which if *it* be supported or fixed, the body or system will remain *in equilibrio* in every position about that centre. Hence the point on which the beam of a pair of scales is poised is its *centre of gravity* or inertia. And the centre of gravity of two bodies, connected by a bar passing through their centres, is determined *by multiplying either body by the whole distance between their centres, and dividing the product by the sum of the bodies, when the quotient will be the distance from the centre of gravity of that body opposite to the one by which the whole distance is multiplied.*

Example. Let the weights A and B, in *fig.* 46., be appended to the beam, 24 feet long. Let A be 4 cwt. and B 7 cwt.: then

Fig. 46.

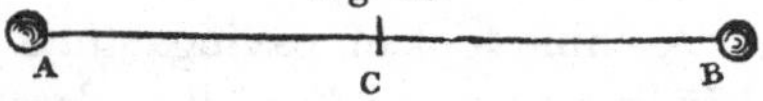

by the Rule $7 \times 24 \div 4 + 7 = 168 \div 11 = 15\frac{3}{11}$ feet, and $4 \times 24 \div 4 + 7 = 8\frac{8}{11}$ feet. But $15\frac{3}{11} + 8\frac{8}{11} = 24$. Hence A and B are respectively $15\frac{3}{11}$ feet and $8\frac{8}{11}$ feet from the centre of gravity of the mass, or, which is the same thing, from the point of equilibrium.

When the weight of the connecting bar is taken into account, and becomes an element of the united mass, we have this

Rule. *To twice the weight of either body add the whole weight of the lever or connecting bar, and multiply the sum by the central distance; then divide the product by twice the mass compounded of the bodies and the bar, and the quotient will be the distance of the centre of gravity from that body opposite to the one whose double is employed in the first step of this process.*

---

* Newton, at the age of twenty-two years, in contemplating and reflecting on the fall of an apple from a tree, led mankind into the belief of the law of universal gravitation.

Example. Thus, if the beam be 26 feet, and weigh 1 cwt. and the bodies be 4 and 7 cwt. respectively, the system will equilibrate if the point of suspension be $16\frac{1}{4}$ feet from the 4 cwt. mass, and $9\frac{3}{4}$ from that of 7 cwt.

Specific gravities are determined by dividing the weight of the body, of the same bulk, by the weight of air or water: air being 1·0000, all the gases are referred to this standard, some being more, others less than unity. So also of vapours, air being 1, all the others are referred to this standard: of liquids, distilled water being 1, all other liquids are compared by it; and the specific gravities of solids are estimated by that standard, the temperature being 64° 1′.

The weight of the atmospheric air at 32° is to distilled water as 1 to 770, and of air to mercury as 1 to 10466.

A cubic inch of zinc or of cast iron weighs 4·16 ounces: of steel and bar iron $4\frac{1}{2}$; brass 4·858; copper 5; silver 6; lead $6\frac{1}{2}$; cast gold $10\frac{1}{3}$th; pure platinum 11·285, and laminated $12\frac{3}{4}$ ounces.

A cubic foot of paving stone weighs 151·lb.; millstone 155 lb.; granite 165·87 lb.; slate 167 lb.; marble $171\frac{1}{8}$ lb.; chalk 174 lb.; basalt 179 lb.; limestone $198\frac{2}{3}$ lb.

A cubic foot of oak is from 54 to 73 lb.; of box 57lb.; yew $50\frac{1}{2}$ lb.; ash $47\frac{1}{2}$ lb.; beech $43\frac{1}{2}$ lb.; walnut 43 lb.; elm $34\frac{3}{4}$ lb.; larch 34 lb.; poplar 24 lb.; cork 15 lb.

A cubic foot of rain-water weighs $62\frac{1}{2}$ lb.; of salt-water $64\frac{1}{2}$ lb.

Cubic Inches in 1 cwt.

	cub. in.		cub. in.
Cast iron -	- 430·25	Cast copper -	- 352·41
Bar iron -	- 397·60	Cast lead -	- 272·8
Cast brass -	- 368·88		

Cubic Feet in 1 ton.

	cub. ft.		cub. ft.
Paving stone -	- 14·835	Honduras mahogany	64·00
Common stone	- 14·222	Mar Forest fir -	- 51·65
Granite -	- 13·505	Beech - -	- 51·4
Marble -	- 13·07	Riga fir - -	- 47·762
Chalk -	- 12·874	Ash and Dantzic oak	47·158
Limestone -	- 11·273	Spanish mahogany -	42·066
		English oak -	- 36·205
Elm -	- 64·46		

In machinery, power is gained by increasing the velocity of the acting force, and, when speed is required in the work, by diminishing the velocity of that force. The contrivances for conveying the power of a machine are wheels, and pinions on axles; conical wheels; rack-work; belts, bands, and chains; single and double cranks; the universal joint; the sun-and-planet wheel; the ball and socket. Levers of particular construction convert circular into rectilinear motion. Spiral gearing is used when the teeth of wheels are cut obliquely, to act with slight friction. Conical drums, inverted to each other, increase or decrease velocity, as desirable; the same end is achieved by wheels of varied diameter on parallel axes, which can be put in or out of gear at pleasure. Excentric crown wheels vary velocity. Tilting hammers, fulling hammers, &c. are worked by cams or wipers, connected with the axis of motion. The toggle joint is a lever of oblique action.

Machines *direct and modify force* transferred to them; and their power is ascertained by dividing the velocity of the *action* by the velocity of the *power*. For the actual effect, multiply this by the *force* of the power, and deduct a *fourth* for friction.

Fly-wheels equalise power: thus a force of 50 lb. per second, imparted to such a wheel, so accumulates as to enable to it overcome 500 lb. in ten seconds.

The *moving* powers are water, air, steam, and animal labour. The *mechanical* powers are the lever, the wheel and axle, the pulley, the inclined plane, the screw, and the wedge.

In *levers,* the power is reciprocally as the lengths on each side of the fulcrum of motion. The power gained in the wheel and axle, is as the radius of the wheel to the axle. In a single moveable pulley, the power gained is double: in a continued combination, it is twice the pulleys less one. On an inclined plane, the power gained is as the length of the plane to the length of the base. The velocity in descending to that falling perpendicularly is as the height to the length, and the force is the same.

Hence a body moving down an inclined plane moves four
times as far in two seconds as in one. The power of the
screw is as the circumference to the distance of the threads,
or 6·2832 that distance. The power of the wedge is as the
length of the two sides to the thickness. The diameter of
the wheel of a pulley should be five times its thickness; the
pin one twelfth; one twelfth allowed on each side for play.

Horse power is reckoned as constant; but the horse can
with difficulty work eight hours a day for any length of
time.

Smeaton reckoned a horse's power equal to that of five
men. Desaguliers says a horse can draw 200 lb. $2\frac{1}{2}$ miles
an hour for 8 hours a day, or 243 lb., 6 hours. Smeaton
reckoned that a horse loaded with 224 lb. could travel 25
miles in 7 or 8 hours. In forming a comparison between
the effects producible from different sources of motion,
the common standard of reference is in this country the
force of a horse which raises 33,000 lb. 1 foot high in a
minute, abstracting every kind of resistance but the load
itself. But the actual medium power of this useful animal
does not exceed 22,000 lb. raised 1 foot high in a minute.
Thus we see how the power of animals is to be estimated
by time; how resistance may be expressed by weight;
and the overcoming that resistance by raising the weight
through a vertical space: it being always supposed that
the weight will remain at the point to which it is raised.
If a single cart-horse draw a ton on a common road at the
rate of $2\frac{1}{2}$ miles an hour, the friction being $\frac{1}{12}$ the load,
the pressure is $\frac{1}{12}$ of 2240 lb.; and the labouring force
$$\frac{2\cdot5 \times 5280 \times 2240}{60 \times 12} = 41,066 \text{ lb.}$$

The labouring force of man is about $\frac{1}{5}$ that of the horse;
though on the tread-mill, and in coal-whipping in the
Thames, and that of labourers, &c. it has been estimated
by Smeaton for 8 hours' work at $1\frac{3}{4}$ million of pounds; by
Desaguliers at 2 millions; and by Coulomb in France still
higher. Dr. Young makes it $3\frac{1}{2}$ millions of pounds one
foot in one day of 10 hours.

The horse draws with the greatest advantage when the line of draught is not level with his chest, but inclines upwards. In working in a circle, as in a threshing machine, the diameter should not be less than 40 feet. In turning a winch a man exerts his strength in different proportions at different parts of the circle: but a fly-wheel equalises this. Most force is exerted in pulling the handle upwards, and least in thrusting it horizontally away. The handles of a winch should be put on at right angles to each other, and not opposite, as they often are.

Wheel carriages are most advantageous when large; and, if mounted on 4 wheels, those in front must be of such a diameter that they may revolve under the body of the waggon in turning it round: and broad wheels are preferred to narrow-tired ones. In large waggons the wheels are made *to dish*, and thus become as it were voussoirs of an arch, possessing great strength, and pressing the road more equally.

Water-mills are ancient and very efficient means of power. When the water running upon a small declivity drives the floats, the wheel is called an *undershot*, — when the water falls from above, it is named an *overshot wheel*. These wheels are from 20 to 40 feet diameter, and their circumference moves from 2 to 5 feet each second of time. The water falls about $\frac{1}{7}$th beyond the top of the overshot wheel. In the *breast-wheel* the fall of water seldom exceeds $\frac{1}{2}$ the height of the wheel. With a small stream of water, engineers use the overshot, but with a large body of water the undershot wheel. Smeaton reckoned that the powers necessary to produce the same effect on an undershot wheel, a breast wheel, and an overshot wheel, must be to each other as the numbers 2·4, 1·75, and 1.

Windmills are of very ancient construction, and of great power; and the velocity of the wind, per *second*, is to the turns of the sails, per *minute*, as 5 to 3. The effect is $\frac{1}{6}$th the force of the wind, or generally equal to 11 horses at a walking wheel. If the wind blows 10 miles an hour, its force is $\frac{1}{2}$ lb. per square foot; at 14 miles, 1 lb.; at 20 miles, 2 lb.; at 25 miles, 3 lb.; at 35 miles, 6 lb.; at 45 miles, 10 lb.; at 60 miles, 17¾ lb.; at 100 miles, 50 lb. nearly. To derive the greatest effect from this force, the sails are inclined to the axis from 72 to 75 degrees; and their tips, which trace in fact the circumference of a wheel, often move 30 miles an hour. The diameter of this circle is sometimes 70 feet; the breadth of the sails from 5 to 6 feet.

In the lever, we have to consider three circumstances: 1. The fulcrum or prop supporting the lever, as an axis of motion; 2. The power employed to raise or support the weight; and 3. The resistance, or weight to be raised.

The power and the weight are supposed to act at right angles to the lever, unless otherwise expressed; and, according to the position of the fulcrum or prop, and the power, with respect to each other, we may have —

1st. The prop between the power and the weight, as in the use of a crowbar; 2dly. The weight may be situate between the prop and the power, as in the rudder of a ship or an oar of a boat; 3dly. The prop may be at one end, and the weight at the other, the power being applied between them; as when a man raises a ladder, or the beautiful mechanism in a watch or a clock, where the power acts near the centre of motion by a pinion, and the resistance to be overcome lies at the circumference, upon the teeth.

In the wheel and axle we have an example of a perpetual lever, as in a capstan. Indeed all windlasses, cranes, mills, windmills, and water-mills are framed on the principle of this machine.

Pulleys are either fixed or moveable; by the former we merely change the direction of a power, as when a man pulls up a stone to the top of a building, in which case, though the man does not move from his place, all his power is concentrated in the stone as it goes up. In the moveable pulley, we double the power, and see, in a stone lifted by this means, the strength of one man on the ground equivalent to that of two men applied in lifting the stone to the top of a building. A pair of blocks with a rope we call a tackle.

Every one knows that he ascends a gentle slope easier than when there is much of acclivity. Inclined planes, as embankments of earth, should always be made a little below the angle of repose, or less than an angle of thirty-five degrees. In haw-haws of very stiff clay, the slope may be forty-five degrees. If the force of traction on a level road be one-twentieth, and the slope of a road one in ten feet, an additional force of one-tenth the load is requisite to draw the carriage up, besides what is required to overcome the friction. Hence the requisite force of traction is $\frac{1}{20} + \frac{1}{10} = \frac{3}{20}$ on the slope. In other words, if 1 horse would do the work on the level, 3 would be requisite to go over the hill. On the gradients or slopes of railways, with a friction of only $\frac{1}{240}$th the load, and the incline 1 in 30, nine times the force required on the level is requisite to overcome the ascent. In great slopes, two engines are used; and in very steep ascents, stationary engines are employed to raise the train.

The screw is nothing more than an inclined plane rolled about a cylinder. If the distance of the centres of two threads of the screw be $\frac{1}{4}$ of an inch, and the radius of the lever by which it is turned round be 24 inches, the circumference of the screw will be nearly 150.8 inches; therefore, if we apply a power of 150 lb. to this lever, the force of the screw will be expressed by this proportion: $\frac{1}{4}$: 150.8 × 150 : 1 : 90,480 lb. $= 40\frac{11}{28}$ tons. We see the great power of the screw exemplified in stamping, in the smith's vice, coining-machinery, and in many domestic agents, as presses. The common corkscrew is merely a screw without a spindle or body. In the combination where the cork is drawn by a second screw or toothed rod and a wheel or pinion, we have an instrument that dispenses with the human power of the deltoid muscle, which the common screw requires in drawing a cork.

The wedge is a great mechanical power, used in splitting wood and rocks, which are rent asunder by the force of a blow that separates the compact body; and, when the vertical angle of the wedge is small, it retains every new position between the resisting forces into which it is driven, and every yielding or separation of the mass is rendered permanent. All nails are wedges made with the greatest economy of their material.

Steam. This power is applied in many ways: to give motion to machinery of all sorts; to move ponderous engines and trains on railways; to navigate ships; to discharge water from mines, &c. Wonderful are the revolutions this new power has created in mining, manufactures, locomotive intercourse by land, and connecting by short spaces of time the ends of the earth one with another. Some years ago the steam-engine was applied to plough land. The medium pressure of the atmosphere and of steam at the boiling heat of water is about equal to a column of 30 inches of quicksilver, or to 34 feet of water, or to 15 lb. on the square inch of surface pressed. But, allowing for friction, &c., the effective pressure may be taken at 12 lb. on the square inch; the working pressure about 10 lb., but usually assumed at from 7 to 9·42 lb.; and this force, multiplied by the number of feet the piston moves in each minute, is the momentum or lifting power each minute. The piston works twice the length of the cylinder at each stroke, and at a maximum in a 9-feet stroke, 14 each minute, travels 252 feet, or with a 6-feet stroke, at 21 each minute, it travels 210 feet. And the lifting power per minute, divided by the power of one horse, determines the number of horses equal to the engine's duty. Steam-engines are divided into different classes, as the atmospheric, high-pressure, and condensing, of which our limits do not admit even short descriptions. Suffice it to say, the horse's power is reckoned as a motive force of 33,000 lb. (*i. e.* 528 cubic feet of water) raised to a height of one foot in a minute.

In complicated machinery, *friction* is $\frac{1}{3}$ or $\frac{2}{7}$ of the force, unless diminished by mechanical means.

As a source of labouring power, water is one of the

most beneficent gifts of the Creator, acting by its own weight only, by its momentum only, by both these combined, or by its pressure. In the former we see its power applied to wheels; in the latter, by its action on a piston. If we have a stream of water 5 feet wide and 2 feet deep, flowing at the rate of 4 miles an hour, and can make it fall a height of 10 feet, its labouring force is 220,000 lb. a minute, or 13,200,000 lbs. an hour. Here is a great force, easily formed in many places of this fair island. Rivers flow with velocities in proportion to the elevation of their sources and volumes of water; but their velocity is retarded by constant obstructions and impediments, though not influenced by friction like water in pipes. The Ganges has only a fall of 4 inches in a mile; the Nile 6 inches in 1000 miles; the bed of the Thames is actually lower at London than below Gravesend, nevertheless its waters all run into the German Ocean. In rivers water ordinarily runs 3 miles an hour; their ordinary declivity is about 4 inches a mile: and this explains why the Rhone, drawing its waters from an elevation of 1000 feet above the level of the Mediterranean, does not pour them out with the velocity of water issuing from the bottom of a reservoir 1000 feet deep, or at the rate of 170 miles an hour. Many methods may be resorted to in order to make the mechanical power of water of great utility, whether it runs only one way, or is subject to the action of tides. Lowell, near Boston, is the seat of very flourishing manufactories, wrought by the water power of a canal, which falls 30 feet in 2500 yards. It is 60 feet wide and 8 feet deep, and affords 1250 cubic feet of water per second, which drives wheels of 30 feet diameter. The mud of great rivers, where they unite with the sea, forms in time deltas, and these little islands in time unite with the mainland and form plains. Thus streams do more in one century than the united labour of millions of men could effect in many ages.

CHAP. VII.

HYDROSTATICS AND HYDRAULICS.

THE concluding paragraph of the last chapter briefly introduces the subjects to be noticed in this.

SECTION I.

HYDROSTATICS.

HYDROSTATICS treats of the equilibrium of fluids, or the principles of the equal distribution of fluid pressure. Hydraulics treats of fluids in motion. Taken conjointly, their theory is denominated Hydrodynamics — a branch of physical science and practical mechanics of the utmost utility.

The general principle of hydrostatics is, that, when a fluid mass, in a state of equilibrium, is subjected to the action of any forces, every particle of the fluid is pressed in all directions; and conversely, when every particle of a fluid is pressed equally in all directions, the whole mass is in a state of equilibrium.

The surface of every fluid at rest, or in a state of equilibrium, is parallel to the horizon, or at right angles to the direction of gravity. The subject of levelling on a grand scale depends on this proposition; for it is evident that two or more places are on a level when they are equally distant from the centre of the earth, and a line which is equally distant from the centre of the earth in all its parts is called *the line of true level;* therefore, because the earth is a sphere, that line must be an arc of the circumference.

If a fluid influenced by the force of gravity is enclosed in a bent tube, or siphon (*fig.* 47.), or in any number of communicating vessels, the fluid will not rest until its surface, in each branch be in the same horizontal plane, and the particles in a quiescent state, at equal perpendicular depths,

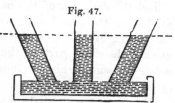

Fig. 47.

are equally pressed.   When a mass of fluid contained in a vessel is in a quiescent state, every particle is pressed in every direction with a force equal to the weight of a column of the fluid, whose base is the particle pressed, and whose altitude is equal to the depth of the particle below the surface; hence the pressure on any particle varies directly as its perpendicular depth beneath the upper surface of the fluid.   The lowest parts of a fluid, therefore, sustain the greatest pressure, and they exert perpendicularly a force equal to the intensity of the super-incumbent mass.   Therefore, the lower parts of vessels containing large masses of water ought to be stronger than the upper.

If we take a cistern whose sides are equal in area to the bottom, the pressure on the four upright sides is equal to twice the pressure on the bottom; but the pressure on the bottom is equal to the weight of the fluid contained in the cistern (supposing it full); therefore, the pressure on the upright surface is equal to twice the weight of the contained fluid; hence, in a cubical vessel, whose bottom is horizontal, the whole pressure on the bottom and the four sides is equal to three times the weight of the fluid which the vessel contains.

Let the box be a cube of 1 foot; then, since a cubic foot of fresh water weighs $62\frac{1}{2}$ lb., the whole pressure on the bottom and three sides is equal to $62.5 \times 3 = 187.5$ lb.

If the vessel be cylindrical, its base horizontal, and its upright surface perpendicular, the pressure on the base is

L

to the pressure on the upright surface as the radius of
the base is to its altitude.

Let the diameter of the base be 3 feet; then, since the
solidity of the vessel is $3^2 \times \cdot7954 \times 6 = 42\cdot9516$ feet, the
whole weight will be $42\cdot9516 \times 62\cdot5 = 2684\cdot475$ lb., being
exactly the fifth part of the weight which measures the
entire pressure, which is therefore equal to $13422\cdot375$ lb.,
or to $5\cdot992$ or nearly 6 tons.

The pressure exerted by a fluid in a quiescent state on
any portion of a vessel, is equal to the weight of a column
of the fluid, having for its base the surface pressed, and
for its altitude the mean depth of the incumbent fluid.

Note. — This mean depth is the same as the distance
of the centre of gravity of that portion below the surface
of the fluid.

But in vessels resembling truncated cones (*figs.* 48.
and 49.), the pressure on the base may be greater or less
than the weight of the
contained fluid, in any
proportion whatever, ac-
cording as the sides of
the vessel converge or
diverge with respect to
the bottom. Hence the
pressure on the bottom

Fig. 48.   Fig. 49.

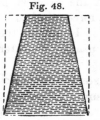
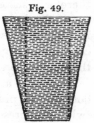

depends solely upon its perpendicular altitude, and not
on the quantity of the fluid; and on this principle any
portion of a fluid, however small, balances any other
portion, however great. Hence the construction of the
hydrostatical bellows and other mechanical instruments,
which, by means of tubes, transmit pressure to the bottom
of cylinders, &c.

The absolute weights of different bodies possessing the
same magnitude are called the specific gravities or densities of the bodies; and any body that, under the same
magnitude, is heavier than another, is said to be specifically
heavier. Hence, if two fluids of different densities in a
state of equilibrum are included in separate branches of a

bent tube, their perpendicular
altitudes above their common
junction vary inversely as their
specific gravities. Thus, in the
annexed sketch (*fig.* 50.), c *a*
and *b a*, are the respective alti-
tudes of the fluids above their common junction, and these
altitudes are inversely as their specific gravities.

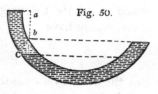

Fig. 50.

Mercury and water are to one another nearly as 1 to
13·6 in weight; therefore to balance a column of water
35 feet high, we have $\frac{35}{13·6}=2·573$ feet. Hence it appears
that a column of water 35 feet high will be kept in equilibrium by a column of mercury 2·573 feet, or 30 876
inches in height.

The converse of the above proposition is also true, that
the pressures on the plane of their common junction are
equal to one another, as the fluids are in a state of
equilibrium.

We must ever recollect that the specific gravity of a
body is the relation of its weight with respect to the weight
of some other body of the same magnitude. And the medium employed is either air, or distilled water at a temperature of 39° of Fahrenheit's thermometer. The density
of water at this temperature being once adopted, and the
weight of a cubic foot of rain or distilled water being 62½
pounds avoirdupois, we have thus a standard of comparison
for weighing, by means of the hydrostatic balance, all sub-
stances which fall under the conditions of the following
proposition.

When the magnitude of a body is given, the density
and specific gravity are directly as the quantity of matter
it contains.

Thus, if two globes M and M', whose diameters are as 4
to 7, and their specific gravities w, w' as 2 to 5, then their
weights stand to each other in the following relation.

We know from mensuration that the magnitudes of the
globes are as the cubes of their diameters; therefore, if

M and M denote the magnitudes, and s and s' the specific gravities, we have

$$\text{M} : \text{M}' :: 4^3 : 7^3 :: 64 : 343$$
$$\text{s} : \text{s}' :: \qquad\qquad 2 : 5$$

But the weight varies as the magnitudes and specific gravity conjointly; therefore, by compounding the above proportions, we obtain,

$$\text{w} : \text{w}' :: 64 \times 2 : 345 \times 5 : 128 : 1725,$$

That is to say, the weight of the globe M, whose specific gravity w is 2, = 128 ; and the weight of the globe M', whose specific gravity w' is 5, = 1725.

In bodies of equal weights their specific gravities are inversely as their magnitudes. This is very evident, for the specific gravities are as the weights directly, and the magnitudes inversely ; consequently,

If a cylinder *a*, in *fig.* 51., of a certain substance, 24 inches high, weighs 10 lb., and it were required to ascertain the height of another cylinder of the same base and weight, but of a different substance, the specific gravities of the materials being as 12 to 1 ;

Since we know their gravities are inversely as their magnitudes, when the weights are given ; it follows, that the magnitudes are inversely as the specific gravities, under the same circumstances.

Fig. 51.

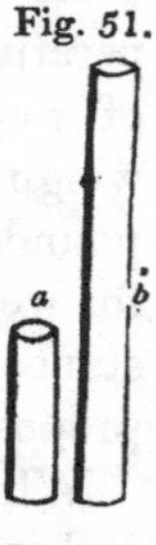

Hence putting *h* and *h'* for their respective heights, we have $h : h' :: 1 : 12$; or $h' = 24 \times 12 = 288$ inches for the height of the cylinder *b*, in *fig.* 51.

Where any body floats upon a fluid, as a ship, or a swan, or is wholly immersed in it, without sinking to the bottom, as a fish, it is pressed upward with a force equal to the weight of the quantity of the fluid displaced, and the direction of that force passes through the centre of gravity of the immersed portion of the floating body.

If the body floating be at rest, the upward and downward pressures are equal one to the other ; that is to say, the weight of the body, and the weight of a quantity of fluid equal in magnitude to the immersed portion, are equal.

A solid body immersed in a fluid of the same specific gravity with itself, remains at rest in all positions. But if the fluid be of greater or less specific gravity, the solid will ascend or descend with a force equal to the difference between the weight of the solid and an equal bulk of the fluid.* And when thus immersed, the weight which the solid loses is to its whole weight as the specific gravity of the fluid is to that of the solid.

This weight is not annihilated, but counterbalanced by a force acting in a contrary direction; hence, in drawing up a bucket of water from a well, we perceive not its weight while in the water, but are sensible of it when it clears the fluid; hence also the strength of dogs in saving persons from drowning.

We see floating bodies take different positions in water; and if the reader experiment with one of those boxes of beautiful solids made by Larkins, he may learn more philosophy of hydrostratics than would fill a book.

The surface of the fluid marks upon the floating body a *line*, called the *water line*, or the *load line*, or the line of flotation. The horizontal sur-
face of the fluid *a b*, in *fig.* 52., is the plane of flotation; and the line *c d*, the line of flotation; also *e f*, is the vertical passing through the centre of gravity of the body *f g h*, and the displaced
fluid space *c f d*. The portion *c d h g*, is the *extant*, and the portion *c f d*, the immersed part of the volume *g f h*. The quantity of fluid displaced by the body is indicated by *c f d*, as a physical line. Thus, if the an-

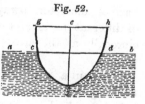

Fig. 52.

---

* We see this exemplified in raising heavy bodies from the bottom of deep waters; carrying ships of burden over shoals or bars by other floating machines called *camels*; in drawing piles that have been driven into the beds of rivers and the sea, &c.

nexed sketch (*fig.* 53.) represent a homogeneous body resembling a parallelopipedon, and *c r, r v,* be each 1 foot,

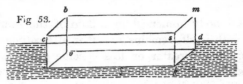

Fig 53.

and *r k,* 10 feet, then the immersed volume *c r v s k d,*
is 10 solid or cubic feet of the body : but a cubic foot of
fresh water weighs 62½ lb.; therefore, the volume of water
displaced weighs 625 lb.; and this is the weight of the
whole body *b r k m.*

<hr>

SECTION II.

HYDRAULICS.

WE proceed now to *fluids in motion,* or *Hydraulics.*

If water flow in a canal or river, or through a pipe of
variable diameter, always filling it, the velocity of the fluid
in different parts of the canal, river, or pipe, will be reciprocally as the areas of the transverse sections in those
parts.

Thus, in a cast-iron pipe (*a b,* in *fig.* 54.) 7 feet long
tapering from 6 inches at one end to 3½ at the other, the
velocity of discharge at the

Fig. 54.

narrow end is thus found.
Supposing it to enter at the
wider end with a velocity
of 2 feet per second, what

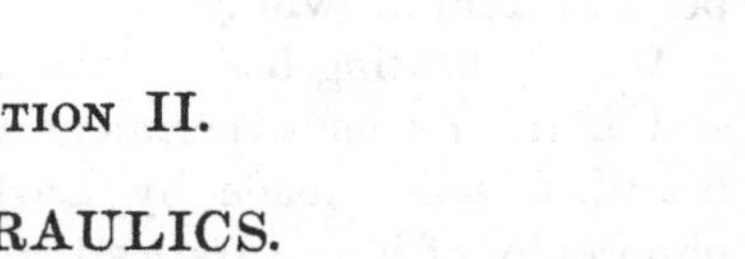

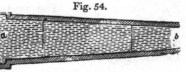

will be its velocity at 3 and 6 feet respectively after entering into the tube.   Here the diameters, estimated in order
at 0, 3, 6, and 9 feet, are respectively 6, 5·16, 4·3, and
3½ inches ; hence

at 3 feet $\frac{961}{36}$: 2 :: 36 : $2\frac{670}{961}$ inches;
at 6 feet $\frac{169}{9}$: 2 :: 36 : $3\frac{141}{169}$ inches;
at 9 feet $\frac{49}{4}$: 2 :: 36 : $5\frac{43}{49}$ inches;

therefore the series of velocities with which the water enters, moves in at fixed points, and leaves the pipe are 2, $2\frac{670}{961}$, $3\frac{141}{169}$, $5\frac{43}{49}$ inches respectively.

When vessels are filled with water, and apertures are made in their bottoms or sides, the fluid issues with a velocity equal to that due to the depth of the orifice beneath the surface of the fluid, or that which a heavy body would acquire by falling from the level of the surface to the level of the orifice. If the vessel be kept constantly full, the quantity of water that issues in one second is equivalent to a column whose base is the area of the orifice, and whose altitude is expressed by the velocity with which the fluid issues.

Many curious problems might be related here about the spouting of fluids, but we must omit these, to notice the phenomena exhibited by the motion of water in pipes, open canals, and rivers. Now for measuring the velocity of rivers, —

Multiply the mean depth of the stream in inches by the declivity in 2 miles in inches; then multiply the square root of the product by 10, and divide by 11 for the velocity in inches per second; or,

Multiply the mean depth of the river in feet by the declivity in one mile in inches; then multiply the square root of the product by the constant number 11·268, and the result will give the velocity in feet per minute.

When the transverse section is rectangular, multiply the breadth of the section by its depth; then divide the product by the breadth plus twice the depth.

Thus, if the breadth be 100 feet, the depth 8 feet, the declivity 3 inches a mile, the velocity will be determined thus : —

$$\frac{800}{100 + (2 \times 8)} = \frac{800}{116} = 6.8965 \text{ feet} = 82.758 \text{ inches; there-}$$

fore by the foregoing rule the velocity is

$\frac{10}{11}\sqrt{82{\cdot}758 \times 6} = 20{\cdot}25$ inches per second, or to $101{\cdot}25$

feet a minute, $= 20{\cdot}25 \times 5$, where $5 = 60'' \div 12$ inches.

In considering the velocity of water flowing through close pipes, of a given diameter and length, with a given head of water, Eytelwein conceives the whole head of water above the point of discharge to be separated into two portions, one of which he supposes to be employed in overcoming the friction and other resistances in the pipe; and the other portion employed in producing the velocity, and forcing the water through the orifice.

The height which is employed in counterbalancing the resistances he considers to be directly proportional to the diameter of the pipe compounded with its length, and inversely as the area of the transverse section, or the square of the diameter, and consequently, on the whole, it varies inversely as the diameter. But the friction varies as the square of the velocity, hence the height equivalent to the friction must vary also as the square of the velocity.

The effect of atmospheric pressure on running liquids is, that, in a tube of considerable length, descending from a reservoir, it quickens greatly the discharge; in fact, it much resembles the operation of a piston. Hence we see in a vessel of water discharging itself by means of a tube in its bottom, a depression of the water surface in the vessel, over the tube; and as the volume of water lessens, this hollow extends itself like a large funnel. In fact, the force of the effluent water diminishes the pressure beneath; on which account the incumbent air, following the stream, finds, as it were, an easier passage, the velocity of the effluent water being always greater in the middle than towards the sides of the aperture, where it is retarded by tenacity and friction.

As regards the friction or resistance of fluids in pipes, an inch tube 200 feet long, placed horizontally, discharges only one-fourth part of the water which escapes by a simple aperture of the same diameter.

The cohesion of the fluid particles is diminished by heat, which, when increased 100 degrees, nearly doubles in certain cases the discharge.

Pumps raise water by the pressure of the atmosphere, and not by suction, as some suppose: they combine both pneumatic and hydraulic principles. By the common pump, water is raised 33 feet above its surface; but practically, we should limit the ascent to 28 feet, at which height the pump will freely act. In the lifting pump a column of water is raised whose base is always equal to the top of the piston, and its height equal to the distance from the piston to the head. The *forcing-pump* is used to convey water further from its bed than either of the other two, which it does by means of a lateral pipe and valve. Fire-engines are two of these pumps in action, to produce a continued stream. The *chain-pump* consists of two square or cylindrical barrels, through which a chain passes, having a great many flat pistons or valves fixed, but moving free of the barrel. There are many forms of pumps, of which a large volume would scarcely suffice to contain the necessary descriptions.

We might have noticed fire-engines, garden-engines, and some other hydraulic machines, but their introduction would swell this article beyond the space it should occupy in this volume.

The ancient water clock of the famous Ctesibius measured time by reason of the uniform discharge of the fluid, in the form of tears, from the eyes of a figure deploring the rapid speed of time; and these tears being received into a suitable vessel, gradually filled it up, and thereby floated another figure that pointed to the hours sketched on a perpendicular scale. This vessel was daily emptied by a siphon, when filled to a certain height, and its discharge, worked by machinery, told the month and the day.

In the sand hour glass, the depth of the volume of this dry fluid does not accelerate the discharge — a remarkable difference — in a simple modification of the same law.

CHAP. VIII.

LAND-SURVEYING.

SECTION I.

DESCRIPTION AND USE OF THE INSTRUMENTS.

Subsect. I. — OF THE CHAIN.

LAND is measured with a chain, called Gunter's chain, from
its inventor, of 4 poles or 22 yards or 66 feet in length.
It consists of 100 equal links; and the length of each link
is, therefore, $\frac{22}{100}$ of a yard, or $\frac{66}{100}$ of a foot, or 7·92
inches.

Land is estimated in acres, roods, and perches. An
acre is equal to 10 square chains, that is, 10 chains in
length, and 1 chain in breadth. Or, it is $220 \times 22 = 4840$
square yards. Or, it is $40 \times 4 = 160$ square poles. Or, it
is $1000 \times 100 = 100,000$ square links; these being all the
same quantity.

Also, an acre is divided into 4 parts, called roods, and a
rood into 40 parts, called perches, which are square poles,
or the square of a pole of $5\frac{1}{2}$ yards long, or the square of
a quarter of a chain, or of 25 links, which is 625 square
links. So that the divisions of land measure will be
thus : —

625 square links	=	1 pole or perch.
40 perches	=	1 rood.
4 roods	=	1 acre.

The length of lines, measured with a chain, are best set
down in links as integers, every chain in length being 100

links; and not in chains and decimals. Therefore, after the content is found it will be in square links; then cut off five of the figures on the right hand for decimals, and the rest will be acres. These decimals are then multiplied by 4 for roods, and the decimals of these again by 40 for perches.

Example. Suppose the length of a rectangular piece of ground be 792 links, and its breadth 385; to find the area in acres, roods, and perches: —

$$
\begin{array}{rr}
792 & 3{\cdot}04920 \\
385 & 4 \\
\hline
3960 & {\cdot}19680 \\
6336 & 40 \\
2376 & \hline \\
\hline & 7{\cdot}87200 \\
3{\cdot}04920 &
\end{array}
$$

Ans. 3 acres, 0 roods, 7 perches.

Subsect. II. — Of the Cross.

The cross consists of two pair of sights set at right angles to each other, upon a staff having a sharp point at the bottom to stick in the ground.

The cross is very useful to measure small and crooked pieces of ground. The method is to measure a base or chief line, usually in the longest direction of the piece, from corner to corner; and while measuring it, finding the places where perpendiculars would fall on this line, from the several corners and bends in the boundary of the piece, with the cross, by fixing it, by trials, on such parts of the line, so that through one pair of the sights both ends of the line may appear, and through the other pair you can perceive the corresponding bends or corners; and then measuring the lengths of the said perpendiculars.

Section II.

THE PRACTICE OF SURVEYING.

Problem I.   To measure a Line or Distance.

To measure a line on the ground with the chain:
Having provided a chain, with ten small arrows, or rods,
to stick one into the ground, as a mark, at the end of
every chain; two persons take hold of the chain, one at
each end of it; and all the ten arrows are taken by one of
them, who goes foremost, and is called the leader; the
other being called the follower, for distinction's sake.

A picket, or station staff, being set up in the direction
of the line to be measured, if there do not appear some
marks naturally in that direction, they measure straight
towards it, the leader fixing down an arrow at the end of
every chain, which the follower always takes up, till all
the ten arrows are used.   They are then all returned to
the leader, to use over again.   And thus the arrows are
changed from the one to the other at every ten chains
length, till the whole line is finished; then the number of
changes of the arrows shows the number of tens, to which
the follower adds the arrows he holds in his hand, and the
number of links of another chain over to the mark or end
of the line.   So, if there have been three changes of the
arrows, and the follower holds six arrows, and the end of
the line cut off forty-five links more ; the whole length of
the line is set down in links thus, 3645.

Problem II.   To survey a Triangular Field, A B C.

$$A P = 794$$
$$A B = 1321$$
$$P C = 826$$

Having set up marks at the corners, which is to be done
in all cases where there are not marks naturally; measure
with the chain from A to P, where a perpendicular would

fall from the angle c, and set up a mark at P, noting down the distance A P. Then complete the distance A B by measuring from P to B. Having set down this measure, return to P, and measure the perpendicular P C. And thus having the base and perpendicular, the area from them is easily found. Or, having the place P of the perpendicular, the triangle is easily constructed.

Or measure all the three sides with the chain, and note them down. From which the content is easily found, or the figure constructed.

PROBLEM III. To measure a Four-sided Field.

A E = 214	D E = 210
A F = 362	B F = 306
A C = 592	

Measure along either of the diagonals, as A C; and either of the two perpendiculars D E, B F, as in the last problem; or else the sides A B, B C, C D, D A. From either of which the figure may be planned and computed as before directed.

PROBLEM IV. To survey any Field of an Irregular Form.

Having set up marks at the corners, where necessary, of the proposed field A B C D E F G (*fig. 55.*), walk over the ground, and consider how it can best be divided into triangles and trapeziums; and measure them separately as in the last two problems. Thus fig. 55., is divided into the two trapeziums A B C G, G D E F, and the triangle G C D. Then, in the first trapezium, beginning at A, measure the diagonal

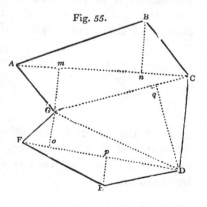

Fig. 55.

A C, and the two perpendiculars G *m*, B *n*.   Then the base
G C, and the perpendicular D *q*.   Lastly, the diagonal D F,
and the two perpendiculars *p* E, *o* G.   All which measures
write against the corresponding parts of a rough figure
drawn to resemble the figure to be surveyed, or set them
down in any other form you choose.

Thus	Thus	The calculation.
A *m* 135	130 *m* G	550 × 180 = 99000
A *n* 410	180 *n* B	550 × 130 = 71500
A C 550		440 × 152 = 66880
C *q* 152	230 *q* D	520 × 120 = 62400
C G 440		520 × 80 = 41600
G F 210	120 *o* G	2)341380
F *o* 175	80 *p* E	1·70690
F *p* 288		4
F D 520		2·82760
		40
Ans. 1A. 2R. 33P.		33·10400

PROBLEM V.   To measure the Offsets.

A *h i k l m n*, in *fig.* 56., being a crooked hedge, or
river, &c., from A measure in a straight direction along
the side of it to B.   And in measuring along this line
A B, observe when you are directly opposite any bends or
corners of the hedge, as at *c, d, e,* &c.; and from thence

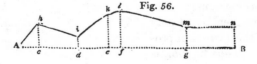

Fig. 56.

measure the perpendicular offsets *c h*, *d i*, &c., with the
offset-staff, if they are not very long, otherwise with the
chain itself, and the work is done.   The register may
be as follows : —

Offs. left.		Base line A B.	
	O		⊙ A
c h	100	80	A c
d·i	50	240	A d
e k	150	370	A e
f l	160	420	A f
g m	90	700	A g
B n	100	890	A B

The spaces included between offsets are calculated as parallelograms: viz., by adding the two perpendiculars together and multiplying this sum by the base; then take the half of the whole when added together for the area; and the work will stand as below:—

$$80 \times (0+100) = 8000$$
$$160 \times (100+50) = 24000$$
$$130 \times (50+150) = 26000$$
$$50 \times (150+160) = 15500$$
$$280 \times (160+90) = 70000$$
$$190 \times (90+100) = 36100$$

$$2)179600$$

$$89800$$
$$4$$

$$3·59200$$
$$40$$

$$23·68000 \qquad Ans.\ 3\text{R}.\ 23\text{P}.$$

When the offsets are long the chain may be used; but short distances are measured with the *offset-staff*, viz., a pole of ten links in length, and each link marked upon it.

PROBLEM VI. To survey an Estate.

If the estate be large, and contain a number of fields, it cannot well be done by surveying all the fields singly, and then putting them together.

1. Walk over the estate two or three times, in order to get a perfect idea of it, until you can carry the map of it

tolerably well in your head; and to help your memory, draw an eye-draught of it on paper, or at least the principal parts of it, to guide you.

2. Choose two or more eminent places on the estate, for stations from which all the principal parts of it can be seen; and let these stations be as far distant from one another as possible.

3. Measure the distances from station to station always in a right line: and in measuring any of these station distances, mark accurately where these lines meet with any hedges, ditches, roads, lanes, paths, &c., &c.; and where any remarkable object is placed, by measuring its distance from the station line, and where a perpendicular from it cuts that line. And thus, as you go along any main station line, take offsets to the ends of all hedges and so on, and noting every thing down.

4. As to the inner parts of the estate, they must be determined in like manner, by new station lines; for, after the main stations are determined and every thing adjoining to them, then the estate must be subdivided into two or three parts by new station lines; taking inner stations at proper places where you can have the best view. And go on thus till you come to single fields.

5. As it is necessary to protract or lay down the work as you proceed in it, you must have a scale of a due length to do it by. To get such a scale, measure the whole length of the estate in chains; then consider how many inches long the map is to be; and from these will be known how many chains you must have in an inch; then make the scale accordingly, or choose one already made.

6. *The Field-book* is ruled into three columns, as shown in *figs.* 57. and 58. In the middle one are set down the distances on the chain-line at which any mark, offset, or other observation is made; and in the right and left hand columns are entered the offsets, and observations made on the right and left hand respectively of the chain-line.

It is of great advantage, both for brevity and perspicuity, to begin at the bottom of the leaf and write upwards; denoting the crossing of fences by lines drawn across the middle column, or only a part of such a line on the right and left opposite the figures, to avoid confusion; and the corners of fields, and other remarkable turns in the fences where offsets are taken to, by lines joining in the manner the fences do, as will be best seen by referring to the Field-book, and comparing it with the plan (*fig.* 59.).

The letter at the beginning of every line is the mark or place measured from; and that at the top the place measured to. Here look at the Field-book with the plan (*fig.* 59.), and you will perceive the first measured line is from ⊙ A to ⊙ B, along a hedge to which offsets are taken at remarkable bends. The 30 links at the beginning means that it is in the same straight line before you come to the station; and at the further end 40 is added beyond the ⊙ up to the road.

7. In computing the contents of estates consisting of many fields, it is usual to make a rough plan of the whole, and from it compute the contents independent of the lines which were taken in surveying. For these, new lines are drawn in the plan, so as to divide them into trapeziums and triangles, the bases and perpendiculars of which are measured on the plan by means of the scale from which it was drawn, and so multiplied together for the contents. In this way the work is very expeditiously done and sufficiently correct; for such dimensions are taken as afford the most easy method of calculation. After all the fields and particular parts are thus computed separately, and added all together into one sum, calculate the whole estate independent of the fields, by dividing it into large and arbitrary triangles and trapeziums, and add these all together. Then if this sum be equal to the former, or nearly so, the work is right; but if the sums have any considerable difference, it is wrong, and must be examined and recomputed till they nearly agree.

8. In all plans of estates it is essential to denote to

M

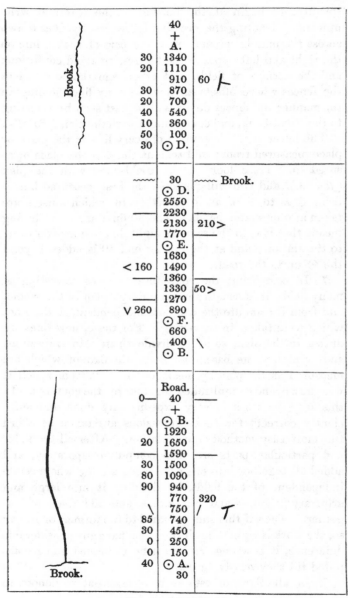

Fig. 57.   *Field-Book.*

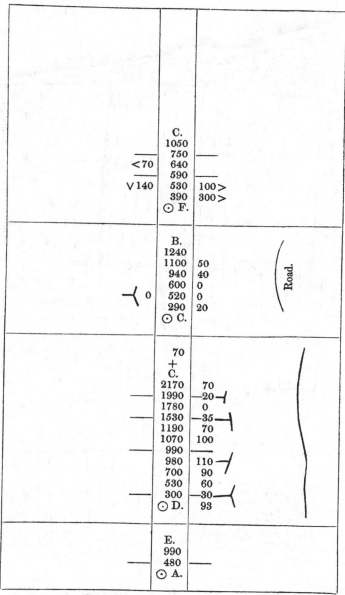

Fig. 58.   *Field-Book.*

M 2

Fig. 59.

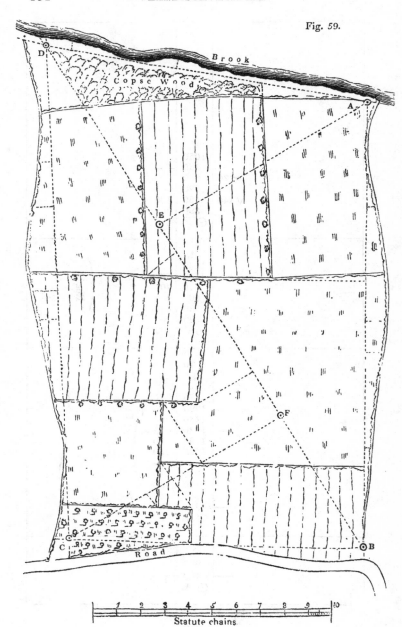

Statute chains.

which fields the fences belong: and, universally, where there is a ditch, that circumstance is defined; the ditch being on the outside of the hedge which bounds the enclosure. Thus, the hedge and ditch A B, in *fig.* 60., belong to the enclosure X; but on the other three sides to adjacent grounds.

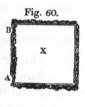

Fig. 60.

PROBLEM VII. To set out small Allotments of Land.

The gardener or bailiff may occasionally be called upon to divide and set out small allotments of land for cottagers to cultivate; and, although sufficient has already been taught to enable the student to accomplish such operations, yet an example or two may help him to manage these matters more readily.

Ex. 1. Suppose a rectangular piece of ground (*fig.* 61.), whose length A B is 800 links, and breadth A C 500 links

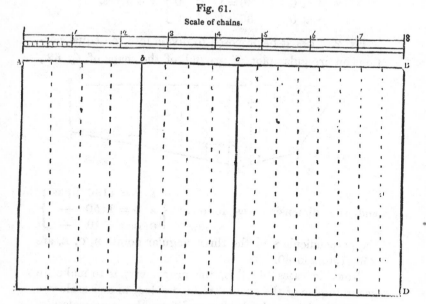

Fig. 61.

Scale of chains.

(4 acres), is to be divided into sixteen allotments, four of which shall contain one rood and ten poles each; four

M 3

more, one rood each; and the remaining eight plots, thirty-five poles each; what will be the dimensions and proportions as laid down to a scale?

Here the first four divisions will
  be found to contain   -   - 125,000 square links.
The second four divisions contain   100,000   —
And the remaining eight divisions 175,000   —

These numbers being divided respectively by the number of links in the line A C; viz. 500, will produce $\left\{ \begin{matrix} 250 \\ 200 \\ 350 \end{matrix} \right\}$

to be set off on the line A B; as A $b=250$, $b\,c=200$, c B $=350$. And these spaces again subdivided will show the proportions of each allotment, which will stand thus:—

	A.	R.	P.		A.	R.	P.	
4 at	0	1	10	= 1	1	0	
4 at	0	1	0	= 1	0	0	$\Big\} = 4$ acres.
8 at	0	0	35	= 1	3	0	

Ex. 2. Let it be proposed to divide into ten equal portions an irregular plot of ground, of the shape of *fig.* 62.,

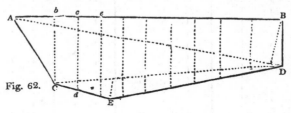

Fig. 62.

and the dimensions as follows: $\left\{ \begin{matrix} \text{A B} = 1130 \text{ links} \\ \text{A D} = 1150 \ — \\ \text{C D} = 940 \ — \end{matrix} \right\}$

the perpendiculars to the three angular points B, C, E, are 250, 190, and 80.

Now, in cases like this, the proper way is to make an accurate plan of the ground to be divided, to any scale you please, the larger the better, and from the same scale plot in each allotment (found as below), which will then be easily transferred from the paper to the ground itself.

Here, from the above dimensions, the area will be found to be 291,000 square links, which being divided by 10, will give the area of each division equal to 29,100 square links.

Let the several allotments run at right angles with the line A B : then, by the application of the scale, c b will be found to measure 300; therefore the half of this, to divide 29,100 by, will produce 194, the distance A b, consequently the triangle A b c is one allotment sought. Now for the next, the average height will be found to be 310; consequently, dividing by this number the area as before, the distance b c will be found to be 97 nearly. Proceeding as before, the average height will by the scale be found to be 330, therefore c e will be found (as near as can be laid down) 88, for the third allotment. And for all the rest, the process is the same. Also, the quantities, reduced from square links, will each be 1 rood 6·56 poles.

In cases where the land varies in quality, and it is desired at the same time that the cottagers should one with another have an allotment of equal value, the intelligent gardener or bailiff will of course give an additional quantity where there is a defect in value. An allowance too should be made for paths which may be common to two or more allotments in setting out, all of which should be properly considered.

Fig. 63.

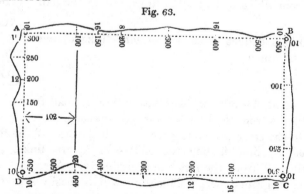

Ex. 3. *Fig.* 63. represents an unequal-sided piece of

M 4

ground, measuring 1·882 acre, which it is required to lay off in five allotments. The first thing to be done is to ascertain the area of the piece. In doing this, let the measurer holding the back end of the chain start from A, noting the offsets on that line in his field-book as nearly as he can to the lines in the figure. When he has marked all the offsets on that line to B, with his cross staff let him set off the line B C, marking the offsets in the same manner. Then set off the line C D, also at right angles. If this has been correctly done, the line D A will also be found to be at right angles with the other lines.

The offsets, being all either triangles or trapezoids, must be calculated by the rules applicable to such figures (as in page 102.): and these, added to the area of the interior rectangle or parallelogram, will be found to amount to 1 acre 88,200 links, which, divided by 5, gives 37,640 square links to each allotment, or 1 rood, 20 poles, 7 yards.

Supposing the lines of division to run in the direction of and parallel to the line A D, we find the offsets on that line to measure - - - links 5600

To the first offset 1500 on the line A B - 1500

With ten times ten links at the corner - 100

And 100 links added to 250 in the line C D - 350

7550

From which deduct 1000 links, taking the half 500

7050

Which deduct from the square links in one allotment - - - - - 37640

30590

and divide by 300, which gives within a small fraction of 102 links from the line A D for that division.

On the same principle proceed with the rest of the four divisions, first finding the offsets in square links, and setting off from the last line accordingly. A rectangular piece of ground or parallelogram will be easily set off in the same manner, without any reference to offsets.

CHAP. IX.

LEVELLING.

LEVELLING is the art of finding a line parallel to the horizon at one or more stations, in order to determine the height of one place with regard to another.

A truly level surface is a segment of a spherical surface, which is concentric with the globe of the earth.

A true line of level is an arc of a great circle concentric with the globe of the earth. Hence, two or more places are on a true level, when they are equally distant from the centre of the earth. Also, one place is higher than another, or out of level with it, when it is further from the centre of the earth; thus, taking the surface of the ocean as an elastic band covering the lower part of the shell of the earth, and yielding to the lunar attraction, we should estimate all heights in reference to this *datum.*

The apparent level is a straight line drawn tangent to an arc or line of true level. Every point of the apparent level, except the point of contact, is higher than the true level. Thus, let E A G, in *fig.* 64., be an arc of a great circle drawn upon the earth. To a person who stands upon the earth at A, the line H D is the apparent level to his rational horizon; but this line,

Fig. 64.

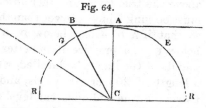

the farther it is extended from his station A, the farther it recedes from the centre; for B C is longer than A C, and D C is longer than B C. Hence, we discover that the line of sight given by the operations of levels, is a tangent, or a right line perpendicular to the semi-diameter of the earth at the point of contact, and rising always higher above the true line of level the further the distance is; this line of sight is very properly called the apparent line of level. The difference, it is evident, is always equal to the excess of the secant of the arc of distance above the radius of the earth.

The common methods of levelling are sufficient for landscape gardening or building, and for conveying water to small distances; but in levelling the bottoms of canals which are to convey water to the distance of many miles, the difference between the apparent and true level must be taken into account. Thus, let I A L, in *fig.* 65., be an arc of a great circle upon the earth. Let it be required to cut a canal whose bottom shall be a true level from A to B of the length of 5078 feet. The most obvious method is to place the levelling instrument in the bottom of the canal at A, and looking through

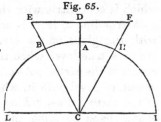

Fig. 65.

the telescope sights at a staff set up perpendicularly at B, to make a mark where the visual ray or point of the apparent level cuts the staff at E; and then to sink the bottom of the canal at B as much below E as A is below D. But this will not give the true level, for it is calculated that at the distance of 5078 feet the apparent level is 7 inches above the true level, and therefore, to make a true level, B must be sunk 7 inches lower than the apparent level directs; so that if A be 4 feet below D, B must be 4 feet 7 inches below E. In practice it is better to take a station in the middle of the line to be levelled, which should be limited to a length of 200 or 300 yards, and then the difference between the true and apparent level need not be attended to, except in cases where the greatest accuracy is required. Hence, we infer that the difference between the true and apparent level, at any distance, may be found, by the well-known property of the circle, to be equal to the square of the distance between the places, divided by the diameter of the earth ; and consequently it is always proportional to the square of the distance. Now the diameter of the earth being nearly 7958 miles, if we first take the distance = 1 mile, then the excess becomes 7·962 inches, or nearly 8 inches, which may be assumed as the height of the apparent above the true level at the distance of one mile, as referred to in figure 65.*

There is sometimes a difference between the true and apparent level as seen through the instrument, caused by the humidity of the atmosphere, which occurs in moist valleys and early in the morning; but this need not be generally attended to.

The instruments used in levelling are, a spirit level, a measuring chain, and a pair of staves.

A spirit level is an instrument which shows the line of level by means of a bubble of air enclosed with some liquor (spirit, or oil of tartar) in a glass tube of an indeterminate length and thickness, whose two ends are hermetically sealed. When the bubble rests at a certain mark made exactly in the middle of the tube, the instrument is level. When it is not level, the bubble will rise to one end. This glass tube is surmounted by a chromatic telescope from 12 to 20 inches long, through which distant objects can be seen. At one end of the telescope is a little tube, containing the eye-glass and a hair horizontally placed in the focus of the object glass. This tube may be drawn out or pushed farther in by a screw for adjusting the telescope to different sights. At the other end of the telescope is placed the object glass. There is a screw for raising or lowering the horizontal hair, and making the line of vision exactly parallel with the glass tube.

The spirit level is usually mounted on a tripod stand, having on the top a ball-and-socket joint, with plates and screws to adjust the instrument to a perfect level. The

* Two thirds of the square of the distance in miles will give in feet the difference between the real and the apparent level. For refraction deduct one seventh.

telescope should turn on this stand to any point of the compass, and still retain its horizontal position.

For engineering works the dumpty and the 20-inch levels are the best, on account of the great power of the telescopes; but for buildings and landscape-gardening smaller instruments will suffice. The measuring chain is the same as that used for surveying, and has been already described.

Levelling-staves are to measure the distance from the ground to the level of the instrument, and are from 10 to 15 feet long, made in 4-feet or 5-feet lengths to render them portable. They vary in form and in the method of joining. The most simple are semicircular, $2\frac{1}{4}$ inches broad, with the face sunk to protect it from injury, as shown in *fig.* 66., and the pieces put together with a common ferule joint, like a fishing-rod. They are marked on the face with feet and decimals in so distinct a manner as to be readable by means of the telescope at a distance of 200 yards. Printed papers for the purpose can be purchased ready for pasting or glueing on the staves; after which they must be coated with varnish to protect them from rain or dirt.

Fig. 66.

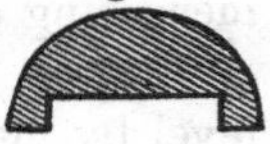

Formerly, when the telescopes attached to spirit levels were so very imperfect that figures on the rods could not be read through them, a sliding vane was attached to the staff to be raised or lowered by the staff-holder at a signal from the operator. This practice was more tedious and liable to error than the present, and is now discontinued.

Two staves are generally employed in taking an extensive line of section.

The operation of levelling is as follows: —

Fig. 67.

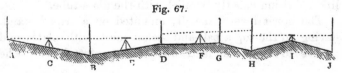

Suppose the height of the point of ground A, in *fig.* 67., above that of the point B, be required. Place the levelling instrument about the middle distance between the two, as at C, and staves at A and B. Set the instrument perfectly level; look through the telescope towards A, when the horizontal hair of your telescope will indicate the point of the staff that is exactly level with the instrument, and the figures the height of that point from the ground; in this case suppose 1 foot. Then turn the levelling instrument horizontally about, that the eye-glass of the telescope may be still next the eye when you look the other way. Ascertain the point of the staff B cut by the horizontal hair, and book the height that point is from the ground, say 9 feet; deduct the former height of 1 foot from it, and the difference, 8 feet, is the true fall of ground from A to B. The horizontal distance between A and B must also be entered in the book, for the purpose of making a drawing of the surface of the ground, which is called a section, and is done by a scale of feet. The same scale may be used for both the horizontal and vertical measurements; or a larger scale for the vertical, according to circumstances. For the purposes of building, or landscape gardening, where the object is to judge of effect, the scales should be alike; but for engineering works, when the quantity of earth which has to be cut or filled in is required to be calculated, the vertical scale is usually made 10 or 12 times larger than the horizontal.

Having thus completed the first set of observations, send the staff A onward to D; set the level at E, and have the staff B kept upon its original site, but turned with its face towards E. Repeat the observations through the telescope; book the height cut on the staves by the horizontal hair, and the distance from staff to staff as before. This will complete the second set of observations; and proceed in like manner, till the whole of the required line of country is passed over.

For the manner of entering observations in your book, write down the heights observed from the level stations in two different columns, viz. in the first column all those observed in looking towards the starting point A, which are called back sights; and in the second column all those observed when looking in the contrary direction, called fore sights. Sum up the heights of each column separately; subtract the less from the greater, and the remainder will be the gross difference of level.

In addition to these, other columns, in which are set down the rise or fall of each observation, are desirable. This is the difference of the height of the back or fore set. If the back set is the greater, it is a rise; and if the fore set, it is a fall; and must be entered accordingly. The difference of the totals of these columns must, if the figuring be done correctly, agree with the difference of the back and fore set columns. Another column for reduced levels is also to be kept. This is simply adding or deducting the rise or fall of each observation from the known height of the preceding station. The difference between the first and last reduced level of each page must accord with the differences of the other columns.

A column for lengths, and one for descriptions or locations, is also requisite.

This is the operation of levelling in its most simple form; but as it is generally requisite that every irregularity of surface of the ground should be measured, and as these frequently occur within short distances of each other, much time would be needlessly consumed in setting the instrument afresh for each; and, therefore, several measurements are taken from the same level-station, which are called intermediate sets, the process of which is as follows: — The instrument is placed as usual, say at F, in *fig.* 67.; the back set D, and fore set G, booked as in the previous example. Now G is to be an intermediate set; so, after having booked it as a fore set, put it down again in the next line as a back set; then send a staff on to H, and book that as the fore set, which completes this double observation, the result of which is precisely the same as if the instrument had been twice placed, once at F, and once between G and H. Any number of intermediate sets may be taken from the same level-station; but they must invariably be booked both as fore and back sets. They will be found particularly useful in crossing water, roads, &c., &c., and may be taken with great rapidity, as extreme accuracy in them is not required, it being obvious from their being entered in both columns that the general levels will not be affected thereby ; but the error will be confined to a misrepresentation of that particular station.

Another species of intermediate set is to book the known height of the instrument (generally 4½ feet) as a station. Thus, place it at I, as shown in *fig.* 67.; book H, as usual; then 4½ feet both as a fore and back set, and J as a fore set. In such a case as this, had the staves been placed at H and J, and I not made a station, the rise of ground at that point would not have been represented, but the ground would appear in the section as straight from H to J. In practice there is no necessity for placing the instrument midway between the staves ; but it is desirable, as, should the instrument be out of adjustment, it will not affect the work, for the error will be equal in the back and fore set, and, therefore, the difference will remain the same.

Neither is it essential that the instrument should be placed on the line of section ; but it may be either to the right or left, as is most convenient for obtaining a view of the staves. Of course the height of the instrument can only be booked as a station when the instrument is on the line of section.

As an example of the real practice of levelling, we give the first portion of the section of the country between Salisbury and Southampton, which was levelled in 1836, for the railway proposed between those places.

The back and fore sets, the lengths and the locations, or descriptions, were entered in the field; the rest of the calculations were made in the evening in-doors. All the columns proved to be correct by the differences balancing, and the work plotted. This will perhaps be better understood by the accompanying page (p. 176.) from the Field-Book.

It will be observed that in the second observation the instrument itself was used as an intermediate station, and is booked in both columns as 4 feet 50.

At the 4th line, 8 feet is booked in both columns as an intermediate station. The same thing occurs in the 6th, 8th, 12th, 13th, 15th, and 16th lines.

Back Set.	Fore Set.	Rise.	Fall.	Reduced Levels. Rise.	Reduced Levels. Fall.	Lengths.	Locations.
8·42				146·37 known height above datum line.		030	Middle of T. P. Road to Southampton.
14·16	4·00	4·42		150·79		070	
4·50	4·50	9·66		160·45		150	
11·18	2·98	1·52		161·97		400	
8·00	8·00	3·18		165·15		990	
1·71	4·45	3·55		168·70		1520	
9·10	9·10		7·39	161·31		1690	
7·81	14·81		5·71	155·60		1790	Farm road.
10·96	10·96		3·15	152·45		1820	Farm yard — Jones.
2·48	11·81		0·85	151·60		1940	
2·09	12·49		10·01	141·59		2160	
2·66	12·41		10·32	131·27		2260	
2·95	2·95		0·29	130·98		2400	Edge of stream 4 ft. deep (Gen. Wyndham).
2·95	2·95		0·00	130·98		3000	Edge of mill-stream 3 ft. 6 in. deep.
8·92	3·83		0·88	130·10		3600	
8·96	8·96		0·04	130·06		3660	Milford Road.
5·18	5·18	3·78		133·84		000	
10·60	1·14	4·04		137·88		000	Working round an obstruction.
12·86	0·58	10·02		147·90		4250	
3·33	8·47	4·39		152·29		4645	Foot-path.
	12·12		8·79	143·50			
138·82	141·69	44·56	47·43	2·87			
2·87	2·87	2·87					

The accompanying diagram (*fig.* 68.) represents this in section plotted to a horizontal scale of ten chains to an

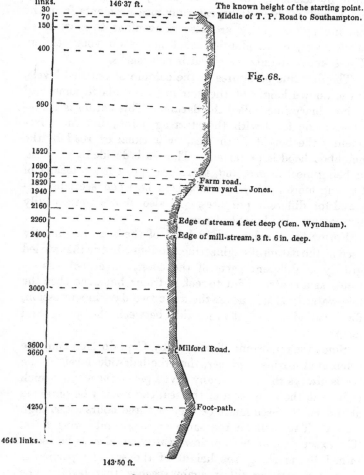

Section of Ground near Salisbury.
Horizontal scale 10 chains to an inch.   Vertical scale 100 feet to an inch.

inch, which is equal to 8 inches to a mile; and a vertical scale of 100 feet to an inch. It is requisite, in making an extensive section, occasionally to take the height of im-

N

movable objects on or near your line of operations, that in case of being compelled to leave your stations, or losing them by accident, you may resume your work from that point, and not have to repeat your measurements from the starting place. They are also convenient to start fresh levels from, or to identify exact points on your section. These known points are called bench marks.

The first line of figures in the column of reduced levels, is the known height of the starting point above some real or imaginary line, called the datum. This datum is sometimes made level with the starting point; but more frequently the height of the sea, or a canal or road in the neighbourhood is preferred. The sea is decidedly the best, as being most in use, and, therefore, presenting means of forming more readily a comparison between sections prepared for different purposes. It also shows more readily the comparative height of rivers.

Formerly, the low-water level of the sea was mostly used as the datum of engineering sections; but as that varied greatly at different parts of the coast, Liverpool was selected as a central point to reckon from; but even then the low-water level was never the same two days in succession, for it varied in height every tide between the spring and neap.

Since the important discovery by Captain Denham, the celebrated marine engineer, that the half-tide level of the sea is always the same at any given place, the mean, which is, indeed, the true level of the sea, can readily be obtained at the coast three hours before, or three hours after, high water. This will be the same at neap and spring tides. The exact time of high or low water can always be ascertained by marking the height of the sea at a particular time, and waiting till it again reaches that level. The time of high or low water was exactly between the period of the two observations. For example, the time when you mark the level of the sea we will suppose to be three o'clock; the water reaches that level again at five; then we know the time of the tide happened at the interval exactly between the two; viz. 4 o'clock.

There is every probability that, in future, all engineering sections and calculations will be reckoned from the mean level of the sea.

The method of adjusting the spirit level when out of order is simple, and such as will readily occur to the operator, and need not be described here; especially as it may be done by the instrument-maker at a very small cost. Still, it is a commendable caution to prove the accuracy of the instrument every morning before beginning work. It is done in the following manner:— Place the two staves as far apart as the telescope will enable you distinctly to read the figures, and the instrument exactly half way between them, as shown in *fig.* 69. In this position you will be able to ascertain the exact difference of level of the points on which the staves rest; for, should the instrument be incorrect, the inaccuracy will be equally great on both sides, and, therefore, will still produce the true result.

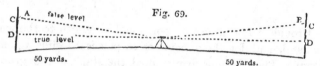

Suppose the telescope to point upwards, say 5 degrees, it will, when directed to the staff A, in *fig.* 69., cut at C, instead of D; and, when turned to the staff B, it will also cut at a similar point C, equally as much above the true level D, as before: the perpendicular in each case being the third side of a triangle, whose two sides and the contained angle, are similar.

Having ascertained the exact difference of level between the two points on which the staves rest, shift the instrument as near to one of them as will allow of your reading the figures. Repeat your observation, and, if the difference between the two staves is the same as the difference in the former instance, the instrument is correct.

As some gardeners may not have a theodolite with a spirit level attached to the telescope, the quadrant described in pages 124, 125, and 204., if carefully constructed, would answer very well for the purposes of the gardener and the farmer. The latter could use it for the construction of water cuts to irrigate his meadows, and for determining the range of his drains, &c. The gardener would find it useful in the formation of terraces, in making ornamental pieces of water, &c., &c.

Ex. 1. Suppose it were required to run a level through the ground indicated by the line A B, in *fig.* 70., from the point A.

Fig. 70.

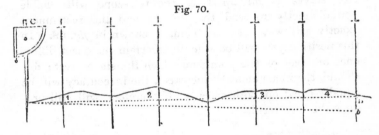

Provide a few staves proportioned in length to the work in hand, and let them have cross pieces to slide up and down. Then having firmly fixed the staff in the ground to which the quadrant is attached, at the point A set the instrument in such a position as the plumb-line shall hang exactly parallel to the perpendicular limb of the quadrant: the upper limb will then be horizontal. This done, direct the eye through the sights, and, at the same time, let an assistant adjust the slides on each staff so as exactly to range with the line of vision. Then suppose the height A C to be five feet; measure five feet downwards from the upper side of the slide upon each staff; so shall the dotted line A B represent the level line required.

Ex. 2. Suppose the operation had been to determine a cut for a drain, to have a fall of 3 inches in every 20 feet. The distance between each staff in the above figure may be supposed to be 20 feet: then, 5 feet 3 inches would have to be measured down the first staff, 5 feet 6 inches down the second, 5 feet 9 inches down the third, &c., &c. The dotted line A *b*, in *fig.* 70., would then represent the line parallel to the bottom of the intended drain.

Ex. 3. If a gardener had to form an ornamental piece of water, the process of finding the level for its bed would be exactly the same as the first example. And if he is pretty well master of what has been taught in the preceding pages, he would have no difficulty in determining to a great nicety the quantity of earth to be excavated, and consequently the cost of the job; for, referring to the figure illustrating these examples, he would only have to calculate with exactness the area of the section marked 1, 2, 3, and 4, and multiply this area by the width of the proposed cut for the cubic content to be excavated.

It was shown in page 171. that the height of the apparent above the true level is (at the distance of one mile) eight inches. Hence the distance or extent of the visible horizon is proportional in leagues to the square root of the observer's height in fathoms; that is, if the heights be 1, 4, 9, 16, &c. fathoms, the distances will be 1, 2, 3, 4, &c. leagues, or 3, 6, 9, 12, &c. miles; or, multiply the height in feet by the constant number 1·5, and extract the square root of the product for the distance in miles. Thus, if the height of the observer be 3262 feet, then $\sqrt{(3262 \times 1·5)} = 69·95$ miles = the distance at which an object can be seen on the horizon. Hence we learn that if a spring be on one side of a hill, and a house on a hill opposite, with a valley between them, and that the spring, seen from the house, appears by a levelling instrument to be on a level with the foundation of the house, at a mile distant, then is the spring eight inches above the true level of the house; and this difference would be barely sufficient for the water to be brought in pipes from the spring to the house, the pipes being laid all the way in the ground.

CHAP. X.

PLANNING AND MAPPING.

PLANNING. — By a ground-plan is to be understood lines representing nothing more than the bases of objects ; such as the space and direction occupied by the foundations of the walls of a house, the fences of a field, &c. The ground-plan of a single tree would be nothing more than a dot, or, if it was a large tree, a small circle. The ground-plan of a box edging is nothing more than a line ; and the ground-plan of a gravel or grass walk is merely a space included between two lines. *Fig.* 71., drawn after the conventional mode, is the ground-plan of an estate, in which merely the fences, roads, watercourses, &c., are indicated.

Mapping. — By a map of an estate is to be understood a drawing or delineation, in which is inserted, not merely the ground-plan of the lines and other objects, but a picture more or less perfect of the superstructure or elevation of these objects.

The first person, as far as we are aware, that made any great improvement on this mode was Mr. Horner, the author of *An improved Method of delineating Estates,* published in 1813, and illustrated by pictorial drawings, partly sold with the work, and partly sold separately, one of the latter being the parish of Clerkenwell. Mr. Horner was a land-surveyor in very considerable practice, but, being a man of genius and of great energy, he occupied himself with a panoramic view of London, and was ultimately brought to ruin, along with a number of other persons, by the erection of the Colosseum for the

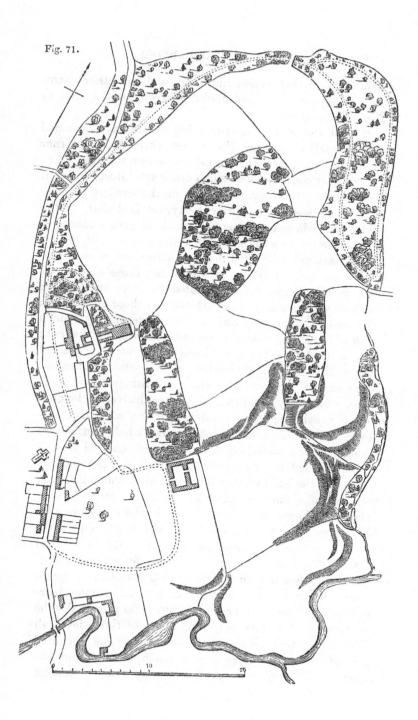

Fig. 71.

exhibition of that panoramic view, and for other extra-
vagant objects.   He died about eighteen months since at
New York, U. S.

" That the art of land-surveying should have remained
almost stationary since the rudest period," Mr. Horner
observes, " is a fact which must excite our surprise, when
we consider the advances which other arts analogous to it
have made.   In comparing some of the earliest delinea-
tions with those of the present day, we find that, in the
former, a rude attempt has been made to give a character
of perspective to what was called the map; trees, water,
and houses, are delineated in them with some faint resem-
blance to nature; while, in the latter, those objects, as
well as all others, are represented by mere emblems or
signs of convention, quite as arbitrary as those of heraldry.
The arts of surveying and of landscape-painting, which
seem to have been united in former ages, are now distinct ;
and, as modern surveyors seldom study the principles of
design, they content themselves with a strict adherence
to precedent, and consider the embellishments of landscape, which they deem extrinsic, as exclusively belonging
to the province of the painter.   Hence, an estate surveyed in the days of Elizabeth, the whole surface of
which has been materially altered by the renewed growth
of timber, and by the general progress of vegetation, as
well as by the hand of improvement, will, under the scale
and pen of a modern surveyor, present a more formal,
naked, dead appearance than it did in the plan drawn by
his predecessor 200 years ago.   Though it abounded with
the most attractive natural beauties, it would be reduced
to a skeleton formed by outlines of the fences and buildings, with a few indistinct though elaborate scratchings to
signify the trees, and a number of parallel waving hair-
strokes of a pen to represent water.   So utterly do the
instruments of a surveyor, like those of time, level all
distinctions, that an estate consisting of one unvaried
tract of enclosures fares as well under them as one en-
riched with every variety of picturesque and romantic

scenery; nay, it fares better, for the delineation of it is not deformed by the rows and patches of stubs and dots introduced into that of its rival, to represent the shady woods and groups of trees which adorn it.   The value of such a production must fall greatly short of the time and toil bestowed upon it, since accuracy is the only criterion for estimating it.

" That a plan may be drawn with the same mathematical precision, and afterwards so finished as to form a faithful and interesting picture of the various features of the property, comprehending the prospects which it commands, as if beheld in a camera-obscura, or from a lofty eminence, has been proved by the enlarged specimens which the author has recently submitted to public inspection.   In these, the whole subject country is represented in the colours of nature, and all its parts are drawn in a correct and faithful manner.   To that portion which represents the estate itself, the scale is universally applicable, while the delineation of the country bordering on it gives a lively idea of the relative bearings of the different parts on each other." (p. 9.)

Notwithstanding " the superior advantages," which Mr. Horner says attend " this style of delineation," we can by no means approve of it, either in point of utility or taste.   In the first place, it cannot be carried into execution by any person who is not, as Mr. Horner himself was, a proficient in drawing landscape — in fact, a drawing-master.   2d. The map of an estate so delineated does not readily admit of marking on it alterations in the fences, roads, or other features, or even of taking dimensions.   3d. In point of taste, the effect is bad of placing the compass, scales, &c., as part of the scenery of the foreground.   On the whole, it seems to us an attempt to join together two things which are incompatible.   A better mode would be, to show the estate in geometrical profile in the centre of the map, and to surround it by a panoramic bird's-eye view of the scenery as far as the horizon on every side.   Such a map we have shown in the *En-*

cyclopædia of Gardening, edit. 1835, fig. 639. p. 631. ;
and which we here repeat (*fig.* 72.).

Fig. 72.

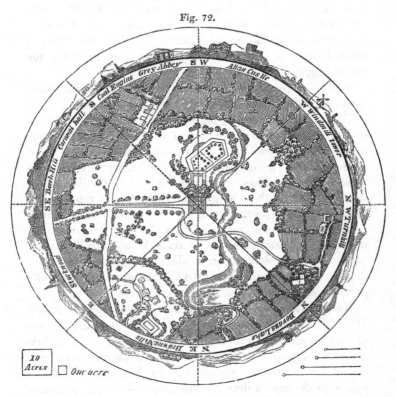

After maturely considering all the various modes in
which estates have been delineated, we are convinced that,
in the present state of our knowledge, the three following
modes are those alone which deserve to be adopted.

The Conventional Mode. The most common is the
conventional mode ; in which the situation of a wood is
indicated by a few scattered trees ; a coppice, by a few
bushes ; hedges, by a simple line or a slight fringe to give
an idea of vegetation ; and buildings, sometimes by a plan
only, and at others by an attempt at an elevation. This
mode of delineating an estate is represented in *fig.* 71.

The Vertical Profile. This mode consists in represent‐ ing every object on an estate as it would be seen by the eye placed immediately over it. It differs from a bird's‐ eye view, inasmuch as, in the latter case, the eye is supposed to be placed directly over the centre of the map, in consequence of which all the objects will diminish as they approach towards its extremities. This mode, it is obvious, could never be adopted in a plan or map that was to become a subject of reference for dimensions or superficial contents. The vertical, or geometrical profile, therefore, is the mode decidedly to be preferred. It does not give a picture of the elevations or sides of objects ; but if we furnish all these objects with shadows, and sup‐ pose the position of the sun in the firmament to be at an angle of 45°, then the shadows, measuring from the centre of every object, will give its exact height, and as far as that shadow extends its equally exact shape. *Fig. 73.* is a specimen of this mode of mapping.

The Isometrical View. By this mode as accurate a ground-plan is obtained as by any of the preceding modes, while at the same time an equally accurate representation of two sides of every object is obtained, the whole being strictly geometrical.*

If two isometrical views of an object or an estate are taken from opposite points, for example, from the two ex‐ tremities of the diagonal of a square house, then the eleva‐ tion of all the four sides of that house will be obtained with perfect accuracy.

The isometrical delineation, therefore, we consider as by far the most perfect mode of representing an estate for purposes of utility, as well as with reference to landscape improvement. It therefore deserves the particular study of the gardener ; and we would strongly recommend him to read very attentively the Chapter on *Isometrical Pro‐ jection and Perspective,* to be found at p. 207.

---

* *Isostates,* Gr. a surveyor ; *Isometrical,* that which hath limits of exact measure.

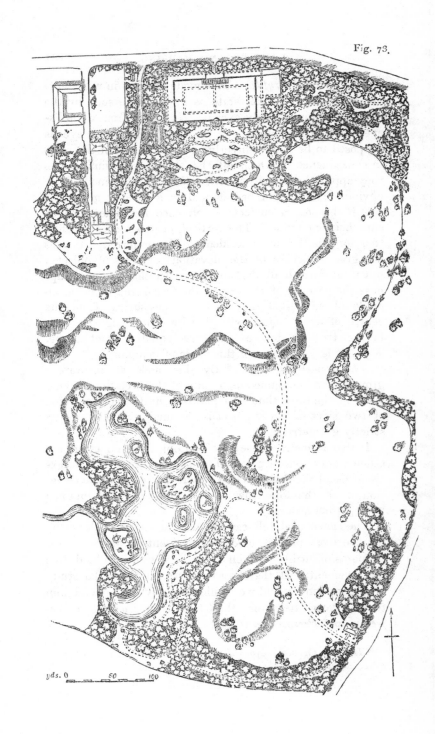

Fig. 73.

yds. 0   50   100

Hints applicable to each of these three Modes. Take care to be particular in placing the ditches of the outside fences with accuracy (see *fig.* 60. p. 165.), because the proprietor on whose side the hedge is placed is considered as its owner, and bound to keep it up as a fence. Indicate the direction of running water, whether in ditches or brooks, by arrows placed in the bed of the brook, or alongside the ditch. Exterior to the plan insert a scale of chains and links, and another of feet and yards, but take care to have these scales quite plain, and not, as they frequently are, made more conspicuous than the lines of the plan. Do not forget to place the north and south point in some spare place outside the plan, with perfect accuracy, and with the north and south line of sufficient length to admit of parallels being taken from it all through the plan. In general, follow the practice of map-engravers, and have the north at the top of the sheet; but deviate from this arrangement when, owing to the shape of the ground mapped, it will be found inconvenient. Place the name of the estate, its proprietor, and the parish and county in which it is situated, with the date of the survey, the purpose for which it was made, and the name of the person who made it, over the centre of the map, at the upper end of the paper; or, if there is not room in the centre, place it at either of the sides. Above all things, let the writing or printing be quite plain, and not overwhelmed with ornamental penmanship, which is to render the means more conspicuous than the end. In the right-hand corner, at the bottom of the sheet, and immediately within the boundary-line, let the party who is responsible for the accuracy of the map, or who has had indicated on it the alterations or improvements which he proposes, sign his name, adding his address and the date. The map, whether a mere survey to show the form and extent of the property, or a plan made to indicate proposed improvements, is now complete.

Shading and colouring Maps. Till within the last fifteen or twenty years, all maps were drawn and shaded with Indian ink or sepia, except such as were executed

on parchment for legal purposes, to accompany title-
deeds, &c. ; these were drawn and shaded with fluid inks
of different colours. which, being stains, sunk into the
parchment, and therefore could not readily be erased like
Indian ink or sepia, which adhere to the surface. Common
writing ink is a very good substitute for these stains where
the maps are not to be coloured. At present, Indian ink
and sepia are only used in plans and maps which are not
coloured, or in making the plan-lines of coloured maps.
The trees, hedges, and other raised objects are outlined
with the black-lead pencil; the colours are applied of the
same shade of depth which it is wished they should present when finished, and the shadows are added by the an-
tagonist or complementary colour, with a little grey added,
especially round the edges of the shadow. For further
details we must refer the reader to what is said on the
shading and colouring of plans in the Chapter on *Architectural Drawing*, p. 204.

The splendid lines of railway in Great Britain, Ireland,
and elsewhere, afford many fine points of view from which
may be collected rich and varied ideas on this novel and
interesting matter. These instructions, so far as they go,
are sufficient in ordinary cases ; but surveying, levelling,
and planning and mapping, sometimes require the combination of geological data with that which is merely topo-
graphical ; and when this is the case, sections of the strata
composing the estate must be exhibited in profile, and the
surface of the plan must be marked with the character of
the strata underneath as far as known by mining or boring.
Hence, as it may be sometimes required to indicate the
mineral qualities of an estate, the young gardener should
labour to gain some knowledge of the stony masses of the
earth ; for rocks, marl, and clay follow one another in
such order, that the rich soil which bears luxuriant corn
is sometimes but a small height above the rock out of
which the action of the atmosphere, during successive
ages, may have contributed to effect the wonderful changes
we behold of organised matters covering the surface.

CHAP. XI.

ARCHITECTURAL DRAWING.

REMARKS. — There is, perhaps, no greater gratification which the mind is capable of receiving through the senses than that which is derived from practising the art of drawing ; and while the exercise of the art of ornamental representation is at once innocent and amusing, and a mark of refined taste and education, it is to architectural and mechanical drawing that we are indebted for that perfection in arts, manufactures, &c., that have contributed to our comfort and luxury, and have rendered mankind more wise and more happy. Indeed, in the present state of society, we could not convey our ideas of those things which have become necessary to our habits and comfort without a knowledge of the art of drawing.

Agents who can make plans of estates, and understand the surveys of others ; gardeners who can draw horticultural structures and their details ; and bailiffs who can show by a drawing how to improve or enlarge a farmery, are much preferred by employers to those who do not possess this knowledge ; and every man, whatever his profession may be, should be able to give some idea, by drawings, of the house he should like to inhabit or wish to build. These considerations should lead every young man to acquire some knowledge of the art of drawing.

Geometrical plans and elevations may at first sight appear difficult, and isometrical projections and perspective still more so ; but the learner has only to make a fair trial, and he will find the difficulties to be overcome fewer than he imagined, and ere long he will perceive the comprehensiveness and utility of geometrical drawing, and delight in the beauty and truth of his radial representations, where —

"Miles are measured in an inch of space."

As a means to this end, the following instructions are given, divested of every technicality, as far as the subject will permit.

Drawing Materials. — The learner must first of all be provided with a drawing board, say about 2 feet 6 inches by 2 feet, perfectly square at the angles, and a T square the length of the board. Observe that, if the board is perfectly square, the drawing will be so likewise; but if not, the drawing will be out of square, however correct the T square may be; therefore the truth of the board is far more essential than the truth of the square. Let even small drawings be made with a large square on a large board, for the sake of firmness. The drawing paper is fixed on the board with flat-headed brass pins made for the purpose; but for fine drawings the paper is stretched on the board as follows: damp it on the back with a sponge and clean water; and after a few minutes, when the wet has distended it, lay it on the board with the damp side down. Turn up about half an inch of the edge of the paper against a flat ruler, then wet the edge of a piece of glue and rub it very hard upon the board along the ruler; press down the edge of the paper upon the glued part, and, with a slip of waste paper over it, rub it down with any smooth hard substance. Fix the ends first and then the sides, and when dry the paper will contract and be perfectly flat on the board. The paper is, of course, cut within the glued part when taken off.

Now, as it is inconvenient to change the square every time a line at right angles is to be drawn, we must have small mahogany set squares or triangles, from 4 to 7 inches long, to place on the drawing square for that purpose; but they serve also for other purposes. Certain angles are frequently required in architectural drawing, and those triangles are made to suit the angles most wanted. *Fig.* 74. is an angle of 45° for drawing all

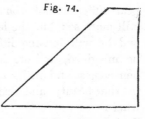

Fig. 74.

right-angled mitres, the sides of an octagon, &c. *Fig.* 75. is an angle of 22½°, which divides the angle of 45° in two, and, when laid with the back against the square, draws the

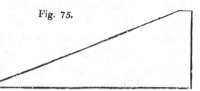

Fig. 75.

pitch of low Italian roofs; it also gives the bevel of window shutters, &c. *Fig.* 76. is an angle of 30°, which, when placed with the back against the square, draws the pitch of low roofs, and in its upright position it is an angle of 60°, and draws the pitch

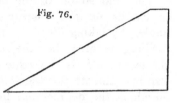

Fig. 76.

of Gothic gables, &c.; it also draws the sides of an equilateral triangle, the lines of a hexagon, &c.; but we shall have more to say of this triangle when we come to isometrical projection. Now a perpendicular line can be drawn by the right angle of any of these set squares (which are here about one third the full size), when placed upright on the drawing square. Professional persons have various other wooden triangles, but we have enough for our purpose. An H pencil, which must be cut with a thin edge, a piece of Indian rubber, and a stick of China ink, complete this part of the materials.

Instruments. — A case of instruments contains a pair of small dividers (compasses) for general purposes, and a pair of large dividers, which is provided with shifting legs for taking large dimensions, and for describing large circles and arcs in pencil or ink; a drawing-pen for drawing straight lines (the curved lines being made with a crowquill); a pair of small bows with pencil foot, and a pair of small bows with ink foot, for describing pencil and ink circles, arcs, and arches. The case likewise contains a scale of equal parts, on which various scales are laid down. (See *fig.* 2. in p. 95.) This instrument has generally the figures 55, 45, 40, 35, 30, 25, 20, at the left hand;

which means that 1 inch makes 55 feet on the first scale, 45 feet on the second, 40 feet on the third, and so on. Observe, if we call the larger divisions 10s, then the smaller divisions on the lower lines are each a foot; but if we call the larger divisions units or 1s, then the small divisions on the upper line of the same scale are each an inch; and in the latter case 40 is a scale of $\frac{1}{4}$ of an inch to a foot, 30 is $\frac{1}{3}$ of an inch to a foot, and 20 is $\frac{1}{2}$ an inch to a foot, and so on of the others. There is also a scale of chords on this instrument for finding the value of angles, by taking 60° from this scale as a radius, and describing an arc; then the chord of this arc, measured on the same scale, is the value of the angle. There are also two instruments for the same purpose, called protractors, on which the degrees are marked, one a brass semicircle, and the other a piece of ivory, like the scale of equal parts, having the degrees radiating to the centre of one side of the instrument; one only of these is necessary. The last instrument to be named (for we think we can easily dispense with all the others) is the sector (see *fig.* 33. in p. 122.), on which are the lines of *sines, chords, tangents, secants,* &c., &c., which have been explained under the head of TRIGONOMETRY; its use in drawing is simply a bevel: having a rule joint, it may be set to any angle, and parallel lines in any direction may be drawn by it, which enables us to dispense with the *shifting*-headed T-square and parallel ruler, the latter being an awkward instrument for a beginner. Before beginning to draw, a few ideas of the Orders of Architecture and their mouldings may be necessary.

Orders. — The five Orders of Architecture, the *Tuscan, Doric, Ionic, Corinthian,* and *Composite,* are the basis on which all classic architecture is founded. Every door, window, arch, moulding, &c. of a structure, bears a certain proportion to the lower diameter of the column of the Order to which the building belongs. Hence the diameter of a column is divided into modules and minutes; there are two modules in a diameter, and the module used for measuring the Tuscan and Doric Orders may be divided into 12 minutes; while, for the Ionic, Corinthian, and Composite, it may be divided into 18 minutes. Each Order has three principal parts: the column, the pedestal on which it stands, and the entablature which it supports; and each part is composed of three divisions; namely, the pedestal has a plinth, dado, and cornice; the column, a base, shaft, and capital; and the entablature, an architrave, frieze, and cornice. In each Order, the height of the pedestal is exactly one third, and the height of the entablature one fourth of the column. The height of the column of the Tuscan Order is 7 diameters of the shaft at its lower part (for columns diminish); the Doric, 8 diameters; the Ionic, 9 diameters; and the Corinthian and Composite, 10 diameters each: consequently, the proportions of the respective principal parts of each order are as follows: — Tuscan Order — pedestals 4 modules 8 min., column 14 modules, entablature 3 modules 6 min.; Doric — pedestal 5 modules 4 min., column 16 modules, entablature 4 modules; Ionic — pedestal 6 modules, column 18 modules, entablature 4 modules 9 min.; Corinthian and Composite each — pedestal 6 modules 12 min., column 20 modules, entablature 5 modules.

With this brief notice of the component parts of the Orders, we shall pass on to the mouldings used in their composition, which must be thoroughly understood, as the mouldings in every building of stone or wood, and in every article of furniture or ornament, have their origin in them.

Mouldings. — The first and simplest is the *fillet*, a square list (*fig.* 77.) which is the smallest member in proportion to the others. Its use is to separate superior members, and to prevent the unharmonious effect that two mouldings composed of portions of circles or ellipses would occasion, when joined together.

The *astragal* (*fig.* 78.) is used as a bead or fillet with a rounded edge, in various compositions, and it is frequently ornamented as at *a*. It is chiefly employed in the Orders,

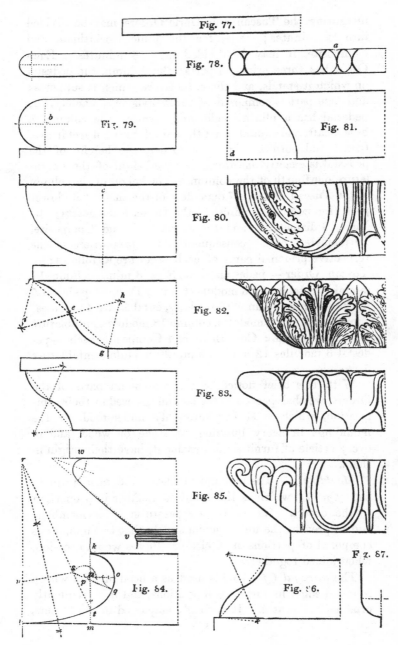

Fig. 77.

Fig. 78.

Fig. 79.

Fig. 81.

Fig. 80.

Fig. 82.

Fig. 83.

Fig. 85.

Fig. 84.

Fig. 86.

Fig. 87.

conjoined with a fillet, in dividing the capital from the shaft of a column; and it is drawn by making the profile, or rounded edge, a semi-circle.

The *torus* (*fig.* 79.) is like the astragal, but of much larger proportion; and, from its rope-like appearance, it seems to bind and strengthen the parts where it is used. Its profile is a semi-circle described from the centre *b*, which projects exactly to a line with the vertical face of the plinth on which the torus rests.

The *ovolo*, or quarter round (*fig.* 80.), is described from the centre *c*. This moulding, from being strong at the extreme parts, is chiefly employed either plain or ornamented, to support other mouldings and members.

The *cavetto* (*fig.* 81.) is described from the centre *d*. It is generally employed in covering and sheltering other members and mouldings, as it is weak in the extreme part, which terminates in a point.

The *cyma recta*, or cymatium (*fig.* 82.), is, from its contour, well adapted for covering other mouldings. It is described in the following manner: — When the projection is ascertained by a scale of modules and minutes, draw the line *f g*, and divide it into two equal parts, as at *e*, which parts will form bases to two equilateral triangles; when their summits, *h* and *i*, will be centres from which to describe the two arcs that join at *e*, and form the cymatium. Ornamented it is very elegant.

The *cyma talon*, or ogee (*fig.* 83.), is drawn much in the same manner as the cyma recta. It is well calculated for giving support, being strong at the extreme parts.

The *scotia* (*fig.* 84.) is employed to strengthen and contrast the effect of other mouldings, and to give a graceful winding to the profile. It is traced in the following manner: — After the points *k* and *l* are ascertained, draw the vertical line *k m*, which line must be divided into three equal parts; then from the point *n*, with the radius *n k*, describe the arc *k o*; next divide the line *o p* into five parts, which will of course make four from *o* to *n*, and one from *n* to *p*; then from *p* draw the arc *o q*, and from *o* describe a segment having for its radius two parts, which will determine the line that passes through the points *q* and *p*. From the point of the plinth, at *l*, raise the perpendicular *l r;* now, from *p* describe another arc having for its radius two parts, this fixes the extremity of the line *q s;* then from *s* describe the arc *q t*, and through the points *t s* draw the line *t r*, making *l u* equal to *t s;* and from the points *s u* raise a perpendicular line until it intersect at *r*, which will be the centre from which to describe the rest of the curve from *t* to *l*.

There are other mouldings, such as the *echinus* (*fig.* 85.), which is frequently ornamented with the egg and spear, but is employed plain in the capital of the Grecian Doric column, between the annulets, *v*, and abacus, *w ;* the *cyma reversa* (*fig.* 86.), which is generally used below the eye in forming one of the mouldings in the bases of pedestals, &c.; and the *congee* (*fig.* 87.), on the shaft of columns.

These mouldings should be drawn on three or four different scales, in order that they may be well impressed on the mind, because a perfect knowledge of their form enables a person to understand the detail of any combination of mouldings.

To draw a Plan. — The disposition of the rooms and their dimensions must first be considered, and a slight pencil sketch by the hand made of them, marking the sizes. This being done, let the labourer's cottage (*fig.* 88.) be the ground plan, which we commence by drawing the front line ; then set off 18 in. from the scale for the thickness of the front wall, and 13 ft. for the inside width of the building, with 18 in. for the back wall. We then take 20 ft. for the inside length, and 18 in. for the thickness of the end walls, which are drawn at right angles to the front and back walls. Make the larger room 11 ft. 7 in., and draw the brick division from front to back, then the shorter division 4 ft. from the front wall. Draw the front door, 2 ft. 10 in. wide, in the centre of the building. Next make the porch, 4 ft. wide, and 2 ft. 4 in. deep, with a small window, 18 in. wide, on either side. Draw the stairs in the

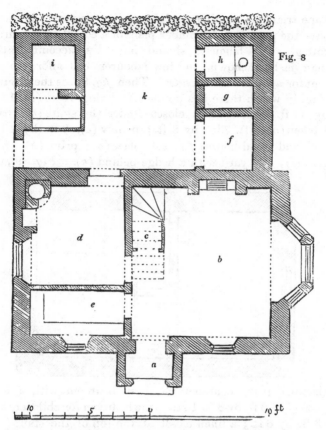

Fig. 8

larger room, with a closet under them, and make the fire-place, 3 ft. wide, in the middle of the space in the back wall. Draw the bevel of the bay-window by the triangle of 45°, and make the middle window 3 ft. wide. Make a 3 ft. wide window, and a 2 ft. 10 in. door, in the back room, also a small fire-place and boiler in the angle as shown: the outbuildings are laid down from the scale in the same manner. The chamber plan (*fig.* 89.) is drawn in the same manner by the scale. The bevels of the end windows in this plan are drawn by the triangle (*fig.* 75.), placed with its back against the square; and the bevels of the two small windows in front of the ground plan are drawn by the

same triangle standing upright on the square. Here we
have the ground and chamber plans of a complete labourer's
cottage. No labourer should have less accommodation
than is here shown, and few labourers can afford to pay
rent for a house much larger.   Then *fig.* 88. is the ground-
plan, in which there is a porch (*a*) ; living room (*b*), 13 ft.
by 11 ft. 7 in., with a closet under the stairs (*c*); back
kitchen (*d*), 8 ft. 7 in. by 8 ft. ; pantry (*e*), 8 ft. by 4 ft. ;
coal and wood-house (*f*) ; ash place (*g*); privy (*h*); pig-
gery (*i*); and yard with a hedge behind (*k*).   *Fig.* 89., the

Fig. 89.

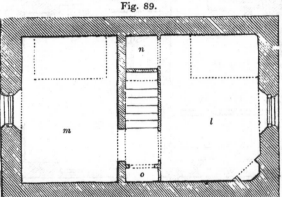

chamber plan, contains the best bedroom with a fire-
place (*l*), 13 ft. by 8 ft. 10 in. ; the bedroom for children (*m*),
13 ft. by 8 ft. ; a linen closet at the top of the stairs (*n*);
and a closet for dresses, &c., over the headway of the
stairs (*o*).

Fig. 90. is the elevation of the front, which is rather
ornamental, but not more so than labourers' cottages
should be, for the effect is produced by very simple means.
In drawing this elevation, we get the horizontal dimensions
from the plans; but for the heights of the easings and
roof, we must make a section, or imagine one, and, by
knowing the heights of the ceilings and pitch of the roof,
we determine the heights of the easings and ridge.   The
heights of the porch and windows, &c., are guided by

what is considered just proportion. We shall have something to say of shading and colouring by and by.

Fig. 90.

Section of a Roof. — A knowledge of the general construction of roofs is highly essential, to show the manner
in which the timbers are disposed and framed, to give the
best bearing.

Fig. 91. is a section of part of the roof of a building
with a para-
pet in which *a*
is the principal
rafter; *b*, the
tie-beam; *c*, the
king-post; *d*,
strutt; *e*, com-
mon rafter or

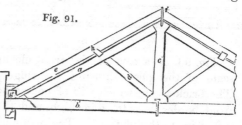

Fig. 91.

spar; *f*, wall-plate; *g*, poll-plate; *h*, purline; and *i*, ridge-board. The foot of the principal rafter and the king-post are mortised in the tie-beam, and for strong roofs they are fixed with straps of iron.

Drawings for a Vinery. — The next nine figures are intended to illustrate the manner of drawing a vinery, roofed on the ridge and furrow principle. The advantage of this mode of roofing is, that the rays from the sun are presented more perpendicularly to the glass in the morning and afternoon, when they are weakest, and more obliquely to the glass at noon, when they are strongest. The dimensions of the vinery must first be well considered, and then the different sizes set off from the scale in the manner in which the ground-plan of the cottage was done. In the present case the length of the range is 80 feet, and the clear width inside 15 feet; the scale will show what the other dimensions are.

Fig. 92. is the plan of a late and early vinery, in which *a a* is the trellis path; *b b*, narrow front path; *c c*, end paths; *d d*, shelves for holding pots containing strawberries, flowers, and other things requiring to be forced, under which are the hot-water pipes; *e e*, pit for general purposes; *f f*, furnaces; and, *g g*, potting sheds.

Fig. 92.

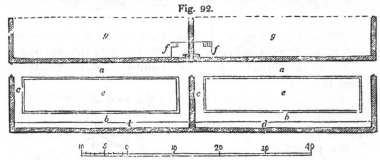

Fig. 93. is a part of the front elevation to double the scale of the plan, showing the ridge and furrow roof. The brick-work, which will of course be covered with the front border, is 2 ft. 6 in. high, and the front lights also 2 ft. 6 in. to the plating. The pediments rise at an angle of 22½°, and the rafters at an angle of 25°.

Fig. 93

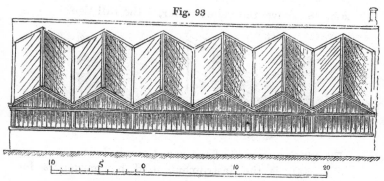

Fig. 94. is the section in which the same letters of reference used for the plan, indicate the same parts.

Fig. 95. is the end elevation.

Fig. 94.   Fig. 95.

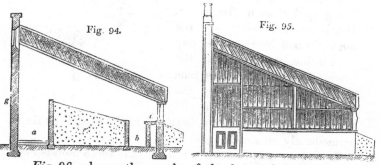

Fig. 96. shows the angle of the bars of the roof, the glass between each bar being in one long sheet.

Fig. 97. shows the ridge rafter with the sash-bar mortised, ¼ the full size.

Fig. 96.   Fig. 97.

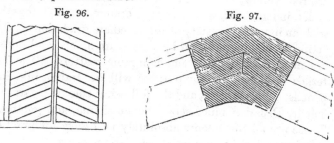

Fig. 98. shows the valley rafter, ¼ the full size.

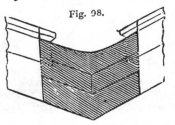

Fig. 98.

Fig. 99.

Fig. 99. is a section of the bar, ½ the full size.

Fig. 100. is the instrument used by gardeners for taking the angles of roofs; it is a mahogany quadrant of about 6 in. radius, with a line and plummet, the quadrant being applied as in the figure. Gardeners generally take the angle formed by the rafter and plumb-line, and call that the angle of the roof; but the true angle is that formed by the rafter and a horizontal line. If the quadrant is numbered both ways, the proper angle is immediately obtained.

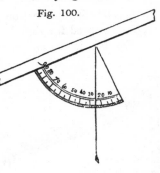

Fig. 100.

It may be only necessary to add, that these houses may be heated either by hot air or hot water, or by both, by having flues from the furnaces, as well as boilers over them.

Shadows, Shading, and Colouring. — The primary colours are red, yellow, and blue; the colours of simple light are violet, indigo, blue, green, yellow, orange, and red, together with an indefinite variety of intermediate gradations. The artificial colours employed by artists, architects, and others, are made from organic and inorganic matters. The materials necessary for our purpose will be Indian ink, hair pencils, colour saucers, and the following colours: carmine, cobalt, Prussian blue, sepia, burnt sienna, and gamboge. These are all the colours absolutely necessary for practical drawings; for, by mixing them, various other colours are produced : for example, carmine and cobalt make a smalt, a neutral tint, a colour for iron, and with Indian ink, an earth colour; carmine and gamboge make a wood colour for sections; Prussian blue and gamboge make an endless variety of greens; and carmine and burnt sienna a mahogany colour, &c. &c.

Shadows. — All shadows in architectural drawings should be thrown at an angle of 45° both horizontally and vertically; that is, the shadow of the head and jambs of a blank window would be of the same breadth, and that breadth would be exactly the depth of the recess. This must be perfectly understood: for instance, a roof projecting 2 ft., and a porch projecting 2 ft. would throw the same breadth of shadow, which would also be 2 ft.; consequently, all shadows thrown on a surface at right angles to the objects throwing them are equal in breadth to the projection; but shadows thrown on oblique surfaces terminate where the angle of 45° from the projection would meet the oblique surface, such as the shadow on a roof from a chimney shaft, &c. Shadows thrown on surfaces parallel to the projection also terminate where the angle of 45° would meet the surface; and all shadows become gradually lighter towards their outer edge. Surfaces may be in *shade* when no shadow is thrown on them; for instance, one cant of the elevation of an octagon would be in shade, while the same part might throw a shadow. All shadows and shades are darkest at the side nearest the light; and they are generally made with Indian ink. These remarks as to the projection of shadows only apply to geometrical drawings.

Application of Colours.—Finished plans are generally tinted with carmine, back-lined, and neatly printed with a crow-quill and China ink, having a scale, but no figured dimensions: but working plans are tinted in the natural colours of the materials used; that is, stone walls are coloured with sepia, brick with carmine, lath and standard partitions, and all other wood-work, with different-toned wood colours. Bearing timbers, or carpenter's work, are tinted darker than the finishing or joiner's work; and sections of timber are tinted much darker than either. A working plan of alterations has all the old walls tinted dark brown or grey; and the new walls, and every part where doors, windows, or fire-places are to be broken out, tinted red. Slight sketches of elevations are touched up with a neutral tint of carmine and blue, or with sepia; but finished elevations are tinted in the natural colours of the materials represented; while working elevations have only a slight tint of blue on the roofs and windows. In tinting geometrical elevations, surfaces that are farthest back are tinted darkest, and projections, &c., in front lightest; but in perspective it is the reverse, for objects are lighter as they recede from the eye, and darkest in the foreground, unless it is a surface on which the light shines. Thus, in shading geometrical figures, the cube, which would have three sides presented to the eye, would be light on the upper surface, a slight dark tint would be laid on one of the other surfaces, and the third would be much darker. So of a prism, or pyramid, and so also of a hexangular prism: for it is manifest that every shadow is a privation or diminution of light by the interposition of an opaque body. By attentively watching the shadows of objects when a brilliant sunshine presents them to our view, the truth of these remarks most forcibly strikes our attention. The same observation applies to shadows produced by artificial light.

South-east is the best aspect for an English house, and south or east the next best. Every cottage should be supplied with abundance of good water, and the drainage around it should be as complete as possible; few things contribute more to the comfort and health of a family than cleanliness within and about the dwelling which it may occupy.

CHAP. XII.

PROJECTION AND PERSPECTIVE.

ISOMETRICAL PROJECTION. — As this method of representation is easier for the learner to practise than perspective *,
we will describe it first; but before doing so, it may be
well just to state the nature of perspective. It is, the art
of delineating on a plane surface the representation of objects, so as to give them the appearance to the eye that
they have in nature; consequently, the rays of light from
the objects *radiate* to the eye, and the objects themselves
diminish in proportion to their distance from the eye, in
the representation as well as the reality. On the other
hand, isometrical projection is the method of drawing the
representation of objects on a plane by *parallel* rays perpendicular to it; consequently, objects do not diminish as
in perspective. Indeed, if the observation be allowed, it
is nothing more than a geometrical elevation upon an angle
which foreshortens the parts of an object; but this angle
must be defined, in order that horizontal and vertical lines
may be measured by the same scale; therefore the given
law is, that the isometrical representation of a cube should
be made by rays parallel to its diagonal, which makes the
three faces seen of equal form and size, and the boundary
line a hexagon. A cube may be correctly projected in any
other position; but we lose the advantage of being able
to measure every part accurately if it is not projected by
rays parallel to the diagonal and perpendicular to the
plane.

Let *a b c d*, in *fig.* 101., be the plan of a cube. Draw up

* *Isometrical Perspective.* This method of drawing was invented
by Professor Farish of Cambridge.

the lines *a l, b k,* and *c i;* then, from the point *f,* with the triangle of 30°, draw *f h* and *f g* till they meet the upright lines. Make *a d e,* an angle of 15°; then *d e* is the height of the cube, which set up upon *f k,* and draw *k i* parallel to *f h, k l* parallel to *f g, l m* parallel to *k i,* and *i m* parallel to *k l.* The cube is now projected, and the three radii and six sides of the hexagon are all of the same length. All the oblique lines are drawn by the triangle of 30° first used one way and then turned over. The scale for measuring this cube would bear the same proportion to the scale of the plan that *d e* does to *d a,* and, if the side of the square were 4 ft., *a d* would be divided into four parts, and *d e* would be divided into four proportional parts. In order to understand more fully the plane of projection, and the rays perpendicular to it, let *p r q s* in *fig.* 102., be the diagonal plane of a cube : then *t r* is the diagonal line of the cube, and *u v* the plane of projection, while *q u* and *s v* are rays; then *u r* and *r v* are equal to any of the lines of the projected cube in *fig.* 101. — Circles are projected by drawing a square round them, and drawing the two diagonals of the square; then the four points where the circle touches the sides of the square, and the four points where it cuts the diagonals, give eight points, which, when projected, will enable the isometrical circle to be completed by hand.

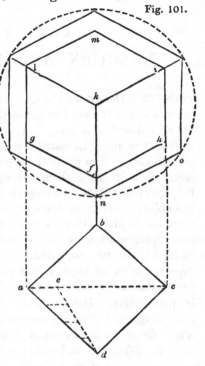

Fig. 101.

Fig. 102.

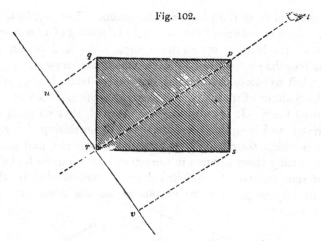

It is obvious that the hexagonal figure 101. is symmetrical with the cell of the bee ; in which the radius is carried six times round the circumference of the circle : and the combination of *these figures* suggests a geometrical combination for tracing on the ground an elegant plan for a flower plot.   For we hold it indispensable, that the young gardener should turn every portion of his knowledge to some account in his daily avocations.

Isometrical drawing is preferable to isometrical projection ; the difference between them being, that the latter is projected in the manner just shown, but the former is simply drawing by the same scale used for the plan.   The two projected cubes in *fig.* 101. show the proportion which isometrical projection bears to isometrical drawing ; the inner cube is the projection, the outer one the drawing, and *n o* in the latter is equal in length to *b c* in the plan.   In isometrical drawing we need not stop to consider the plane of projection or the rays, but proceed to work at once with the double triangle of 30°, *fig.* 103., which draws the oblique lines right and left.   Take every

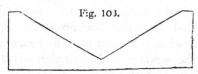

Fig. 103.

dimension from the plan and elevations. Let *fig.* 104. be
the block or outline of the isometrical drawing of a farmery ;
set all the heights up on the upright lines, and set off all
the lengths on the oblique lines, which are drawn to right
and left as before observed by the double triangle, *fig.* 103.
The manner of finding the gables and roofs is shown by the
dotted lines. It is only necessary to add, that all oblique,
curved, and irregular lines are found by enclosing the surfaces which they bound by reticulated squares, and then,
by putting these squares in isometrical drawing, to find the
different points. Isometrical drawings are shaded in the
same way as geometrical elevations on the angle, or, as

Fig. 104.

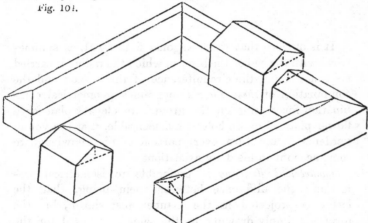

has been said respecting the shading of an octagonal elevation. (See p. 205.)

Perspective. — The objects seen through a pane of glass
in a window, if traced upon the glass, will give some idea
of a perspective representation ; but the use of perspective
is, to enable us to draw the representation of objects on any
opaque surface, without the assistance of the glass, that
shall have the same proportion in form and outline as the
originals. But we must here mention the terms made use of :
the *perspective plane* or picture is the paper or other surface
on which the drawing is made ; the *horizontal line* is a line

drawn across the paper at the height of the eye (by the scale) from the base or ground line. If the eye be raised, the horizontal line is raised accordingly, and the contrary, if it be lowered. The *point of sight* is the place of the eye of the observer; the *distance of the picture* is the distance of the eye from the picture; and the *point of distance* is simply the distance of the eye transferred to the horizontal line for the convenience of drawing. The *centre* of the picture is a point where a perpendicular from the eye would meet the picture; this point need not be in the centre of the paper, but it must always be in some part of the horizontal line. *Vanishing points* are points on the horizontal line and in the plane of the picture, where all parallel lines would meet, except those that are parallel to this plane; these have no vanishing points, but are parallel to each other; and all right lines perpendicular to the plane vanish in the centre of the picture.

To draw *Four Squares* lying on the ground, and receding from the view. Let A B (*fig.* 105.) be the base line, s the

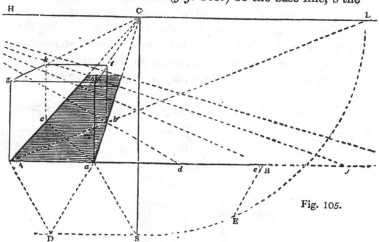

Fig. 105.

point of sight; H L, the horizontal line, and C, the centre of the picture. It will be perceived that the length of s C, is the distance of the picture, and is transferred to the horizontal line, making L the point of distance; and it must be also understood that if A D and B E were produced, they would meet in a point that would be the same distance from the base line that s is from c, — that is, the distance of the picture. Now to draw the squares: — Make A *a* one side of one square, and draw A C and *a* C, to the centre. Draw A L to the point of distance, cutting *a* C in *b;* then draw *b c* parallel to A *a*, and the first square is complete. To draw the other squares, set off on the base line the length of the side of the square, as many times as there are squares to draw, such as *d e f,* &c.; then draw *d e f,* &c. to a point of distance on the horizontal line as far to the left of c, as L is to the right of c (the page is too narrow to show this point), and where these lines cut the radiating lines to the centre c, draw lines parallel to the base, and the four squares are complete. Now to represent a cube on the base line, let the first square be its base, and draw up the perpendicular lines from the four angles. Make A *a*, *h*, *g*, a true square, and draw *g* C and *h* C; then draw *i k*, parallel to the base, and the cube is complete. This diagram shows the method of parallel perspective; the horizontal squares will explain the principle of drawing the square beds of a garden; while the upright cube shows the manner of representing rectangular objects standing above the ground. We have here only an object on one side of the centre of the picture.

To represent Two Cubes, as in *fig.* 106. The same capital

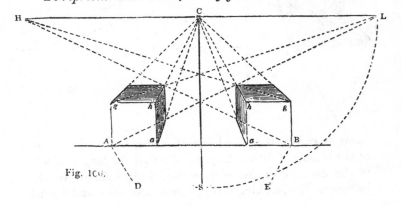

Fig. 106.

letters refer to the same parts as in *fig.* 105.: A B the base line, S the point of sight, H L the horizontal line, and C the centre of the picture. The points at H and L are points of distance; they are the distance of S from the centre; and if A D and B E were produced, they would meet in a point the same distance from the base line as the length of S C; indeed, it is this point which gives the length of S C; it is the point of sight in the plan, and may be at any distance; but whatever the distance is, S C must be the same, and consequently C L and C H, the same. But we shall understand this better presently when we come to the next figure. In the present figure make the two front faces true squares, and draw lines from the angles *a h g* and *a h g,* to the centre. Draw A L and B H to the points of distance, which will determine the depths of the cubes, as in the last figure, when they will be easily completed. The most convenient way to draw an object, say a house for instance, is to make a block plan of it, when we can do so.

To draw a House in angular Perspective. — Artists cannot have plans of the objects they represent, but architects &c. generally have; therefore let *fig.* 107. be the ground plan of a small house laid down by the scale.

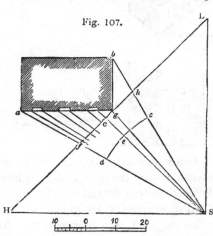

Fig. 107.

Let S be the point of sight, and H L the plane of projection: draw the extreme rays *a* S, and *b* S; bisect these rays with the arc *c d,* to find the centre of the angle in *e;* then draw the perpendicular S C, through *e,* and C is the centre of the picture, while *g* is the corner of the building. Draw S H parallel to the front of the house until it cuts the plane of projection, and s l parallel to the end of the house until it also cuts the plane; then the points of intersection at h and l are the vanishing points. (If the building were not square, then a line parallel to the oblique front produced from the point of sight till it cut the plane would give its vanishing point.) Draw the rays from the door and windows towards the point of sight until they cut the plane; then $f g$ is the fore-shortened front, and $g h$ the fore-shortened end. Now to set up the elevation, draw the horizontal line (about 6 feet above the ground line, *fig.* 109.), and transfer all the lines cutting the plane of *fig.* 107., to the horizontal line of *fig.* 109. Make g l and g h in the elevation the same length as g l and g h in the plan: then h and l are the vanishing points. Set up all the heights from the scale on the corner line g, and draw the horizontal lines to

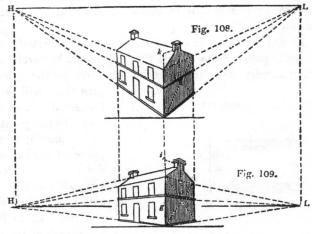

Fig. 108.

Fig. 109.

the vanishing points. Draw the diagonals of the gable, which gives the centre as shown by the upright dotted line; set up the height of the roof on that dotted line to i, then draw from i towards l till it cuts the centre gable line, and from this point draw the ridge of roof towards h. All heights are set up on the corner line, and it will be easily seen how the windows and chimneys are found. *Fig.* 108. is another method of representing the same building, when the eye, and consequently the horizontal line, is considerably above it: the lines at *h* show the manner of finding the height of the roof. The point of sight, s in *fig.* 107., may be in any place relative to the plan, so that the distance from the picture is not less than the length of the picture.   The custom is to choose a point from which the object will look best, and drawings viewed at small angles are more pleasant than those viewed at great angles, with the point of sight too near.   When the fronts of objects are parallel to the plane and above the eye, the representation is called *parallel perspective*.   When the objects are parallel to the plane, but below the eye (as *fig.* 106.) it is called parallel bird's-eye perspective.   When objects are oblique to the plane (as *fig.* 109.), it is angular perspective; and when oblique to the plane, but below the eye (as *fig.* 108.), it is angular bird's-eye perspective. These are merely conditional terms, for there are, strictly speaking, only two kinds of perspective, linear and aerial. Linear perspective is the perspective of lines which we have been describing; and aerial perspective is the art of giving the due tone of light, shadow, and colour of objects according to their distance and the medium through which they are seen.

To draw a Circle in Perspective.—Let *fig.* 110. *a* be the circle ;   draw a square about it, then draw the two diagonals of the square, and where they cut the circle let those be the angles of an inner square. Draw also the two diameters of the circle.

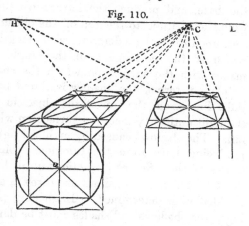

Fig. 110.

Let H L be the horizontal line, c the centre, and the point at H, the point of distance; then put all the straight lines in perspective, which will give eight points of the circle, when the curve is completed by hand.

Shadows and Tinting. The light may be in front, at either side, or even at the back of a picture; but for perspective views of buildings one side is preferable. Of course, all shadows are produced by opaque objects intercepting the light; which may either be natural light from the sun, &c., or artificial light from a candle or other luminous body. The rays of natural light, in consequence of the immense distance of the sun, are parallel; but the rays of artificial light diverge from a point; consequently, shadows thrown by objects intercepting the sun's light are of the same breadth, while shadows thrown by objects intercepting the rays of artificial light increase in width as it leaves the objects. In geometrical drawing the rays of light fall upon the object at an angle of 45°, as we have seen, but the shadows in a picture or perspective drawing are found thus: — We first determine upon the direction in which we wish the light to come into the picture; then find the altitude of the light, which may be high or low according to what we wish the drawing to appear. From the altitude we drop a perpendicular to the base, and produce the latter: we then draw a line from the altitude to the top of the object, and that line continued to the base determines the length of the shadow, which is the co-tangent of the angle that the light makes with the surface on which the shadow is thrown. In plainer terms, the shadow is found by producing parallel lines from the top of the object in the direction of the light until they meet the surface on which the shadows fall. The shadows caused by artificial light are found by producing lines from the luminous point touching the angles of the different projections until they meet the surfaces on which the shadows are thrown, and the length of the shadow is determined. In shading perspective drawings, the shadows and shades must be darkest where they are nearest the foreground ; and each individual part must not have its own light and shadow, but must be blended with the whole.   This is one distinction between the manner of treating geometrical and perspective drawings.   In the foreground of the latter, bright lights are opposed to dark shadows, while the back-ground is blended into one less definite mass.   The same holds good as to tinting — bright and distinct colours are opposed to each other in the foreground, but in the distance they are subdued, as if a mist or haze had come over the objects, which in nature is caused by the intervening body of air.   The eye calculates more readily the size of objects by this gradation of tint, than by the magnitude of the objects in the representation. The proper distribution of light and shadow in a picture can only be properly managed by one who has the eye of an artist ; but any one who can shade polygons, when geometrically represented, will be able to shade and tint a building in perspective.

Reflection and Refraction. — All objects in nature reflect light through the medium of the air, even when the sun is not shining on them ; but it is with the reflection of objects in water that we have to do at present.   It is a universal law of Optics that the angles of incidence and reflection are equal.   The angle of incidence is the angle at which objects are presented to a reflecting surface, and the angle of reflection is the angle at which the image is reflected ; both are equal, but in opposite directions.   Let

$b\,a\,c$, *fig.* 111., be the angle of incidence, then $d\,a\,e$ is the angle of reflection, and $d\,a$ in this case would be the line in which the eye of an observer

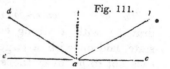

Fig. 111.

was situated ; but images in water will be better understood by a figure.   Let f, g, h (*fig.* 112.) be three trees at different distances from the edge of a piece of water, and at different distances from the eye of an observer on the opposite bank of the water at i ; then k would be the reflection of f ; l the reflection of g ; and m the reflection

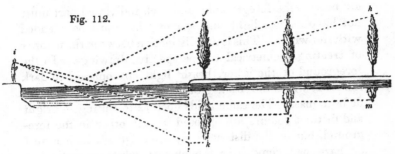

Fig. 112.

of *h*, as seen by this observer. When objects stand on the brink of water, their whole height is reflected; but when they stand at a distance from it, the height of the ground, or rather the height which the ground subtends at the eye, must be deducted from the height of the object for the depth of the reflection. Little need be said of refraction; it simply means the turning away of rays of light from their direct course. If light falls perpendicularly on a piece of water, it passes through it perpendicularly; but if it falls obliquely, then it does not continue the same line of obliquity through the water, but in a direction approaching more to the perpendicular, because water is more dense than air; and if passing through glass, it would be still more refracted from its original direction, as glass is more dense than water; but when it has passed through either body it resumes its original direction in passing through its original medium.

To varnish water-coloured Drawings. — Prepare the drawing with a strong coating of isinglass in a liquid state, laid on with a large flat camel's-hair brush, taking care not to go over the same place twice while wet, for fear of raising the colours. The isinglass must float in a body during the operation, which must be rapid, but with a light touch, lest the colours should run. The drawing thus prepared to receive, when perfectly dry, two or three coats of Masters's patent varnish, each coat being perfectly dry before the succeeding coat is laid on.

CHAP. XIII.

MISCELLANEOUS TABLES.

SECTION I.

INTEREST AND ANNUITY TABLES.

Explanation of the Tables referring to Interest and Annuities.

TABLES I. and II. The *present worth* of a sum of money, as 1*l*., to be received at a future period, is that which, laid out and improved at a given rate of interest during that period, will amount to the proposed sum by the time it becomes due. The difference between the present worth and the sum itself is called the *discount*. Hence, to find the present worth of a sum of money, improved at a given rate for a fixed time, *divide the proposed sum by the amount of* 1*l*. *improved at the assigned rate for the given time, and the quotient will be the present worth : subtract the present worth from the sum proposed and the remainder will be the discount.*

Ex. If 1*l*. be divided by 1·04*l*. the quotient is ·961538*l*. which is the present value of 1*l*. due one year hence ; and, if from 1*l*. we take that quotient, the difference ·038462*l*. is the *discount :* for, ·961538 + ·038462 = 1*l*. Hence the use and application of TABLES I. and II., and particularly the latter, as shown in example 2. page 44.

TABLE III. This table is fully explained in pages 36. and 37., article *Compound Interest.*

TABLE IV. This table is explained, and its construction fully shown, in pages 38. and 39.

TABLE V. The construction of this table is shown in pages 39. and 40. ; and example 1. page 40. and example 5. page 41. show its application and use. Thus, in column 1. the term of years is found, and the reader traces from 21 years across the line to the third column or 5 per cent, and there finds the amount of 1*l.* forborne and improved at Compound Interest to be 12·82153*l.*, which, multiplied by the improved rent of 100*l.*, gives 1282*l.* 3*s.* 1*d.* as the present value of the lease. And, if the reader will substitute 4 per cent for 3½ in the example in page 43., where *Bartering Interests in Property* is treated, and work the question at 4 per cent interest, then 600(28·57142 — 9·38507) =600 × 19·18635 = 11,511*l.* 16*s.* 2½*d.* would be the present value of A's deferred annuity. And the perpetual annuity which A can purchase with 11,511*l.* 16*s.* 2½*d.* is found by a similar proportion to that in the question to be 402*l.* 18*s.* 3*d.*; for 28·57142*l.* : 1 : 11,511·81*l.* : 402*l.* 18*s.* 3*d.*, which is the perpetual annuity A can purchase with 11,511*l.* 16*s.* 2*d.*

But there is another application of this table to which we direct the reader's attention : we allude to the renewal of leases.

Thus, if it be required to find the present value of 100*l.* a yearly rental for 7 years, to commence at the end of 14 years; we have, at 5 per cent. interest in Table V., and opposite 21 years, the present value of 1*l.* per annum

given as	- -	- £12·821153, and
of 14 years equal to	-	- 9·898641
and the difference	-	- £2·922512
multiplied by	-	- 100
gives as the present value		£292·2512, or

292*l.* 5*s.* 0*d.* which a party would now pay for the extension of his lease to 21 years.

Proof: — 292*l.* 5*s.* 0*d.* improved at compound interest for 14 years amounts to 578*l.* 6*s.* 3*d.*; for it is 292·251 × 1·979931 (found in Table III. opposite to 14 years), and 578·3125 × 1·4071 (opposite 7 years and under 5 per cent., Table III.) amounts to 814*l.* in 21 years, which is what the rental of 100*l.* a year for 7 years would have amounted to in 7 years, if the landlord improved at 5 per cent. per annum compound interest that yearly rental which in Table IV. and opposite to 7 years is found to be 8·142 × 100 = 814*l.*

TABLE I. *Showing the Interest and Amount of* 1l. *in* 12, 9, 6, *or* 3 *Months:—also, the Present Worth and Discount of* 1l. *due* 12, 9, 6, *or* 3 *Months hence, Simple Interest.* *

Rate of Annual Interest.	Time.	Interest.	Amount.	Present Worth.	Discount.
3 per cent	1 year	·03	1·03	·970874	·029126
	¾ —	·0225	1·0225	·977995	·022005
	½ —	·015	1·015	·985222	·014778
	¼ —	·0075	1·0075	·992556	·007444
4 per cent	1 year	·04	1·04	·961538	·038462
	¾ —	·03	1·03	·970874	·029126
	½ —	·02	1·02	·980392	·019608
	¼ —	·01	1·01	·990099	·009901
5 per cent	1 year	·05	1·05	·952381	·047619
	¾ —	·0375	1·0375	·963856	·036144
	½ —	·025	1·025	·975610	·024390
	¼ —	·0125	1·0125	·987654	·012346
6 per cent	1 year	·06	1·04	·943396	·056604
	¾ —	·045	1·045	·956938	·043062
	½ —	·03	1·03	·970874	·029126
	¼ —	·015	1·015	·985222	·014778
7 per cent	1 year	·07	1·07	·934579	·065421
	¾ —	·0525	1·0525	·950119	·049881
	½ —	·035	1·035	·966184	·033816
	¼ —	·0175	1·0175	·982801	·017199

* In mercantile transactions, the *interest* is usually taken for the *discount:* but this is incorrect, and of doubtful legality. Thus, at 5 per cent per annum, the interest of 1l. is 1s.; and the discount upon a payment of ready money for a debt due one year hence is 11¼d. $\frac{7}{10}$. If we have to deal with large sums, this difference becomes very considerable: thus, the interest of 1000l. for one year at 5 per cent per annum is 50l.: but, if discount be allowed on 1000l. paid one year before it becomes due, the amount is 47l. 12s. 4d. It is manifest, therefore, that if a banker charge 50l. for discounting a bill of 1000l. payable one year hence, in place of 47l. 12s. 4d., he charges 2l. 7s. 8d. more than he is entitled to demand. Moreover, the fifth column of this Table, headed *Present Worth*, shows by inspection how all discounts should be treated; for ·952381 × 1000=952l. 7s. 8d. is the sum the banker ought to pay, and not 950l., on discounting the bill of 1000l, as 47l. 12s. 4d. if improved, at 5 per cent for one year, will amount to 50l.

TABLE II.   *Showing the Present Value of £1 to be received at the End of any Number of Years not exceeding 50.*

Yrs.	4 per Cent.	5 per Cent.	6 per Cent.	7 per Cent.	Yrs.
1	·961538	·952381	·943396	·934579	1
2	·924556	·907029	·889996	·873439	2
3	·888996	·863838	·839619	·816298	3
4	·854804	·822702	·792094	·762895	4
5	·821927	·783526	·747258	·712986	5
6	·790315	·746215	·704961	·666342	6
7	·759918	·710681	·665057	·622750	7
8	·730690	·676839	·627412	·582009	8
9	·702587	·644609	·591898	·543934	9
10	·675564	·613913	·558395	·508349	10
11	·649581	·584679	·526788	·475093	11
12	·624597	·556837	·496969	·444012	12
13	·600574	·530321	·468839	·414964	13
14	·577475	·505068	·442301	·387817	14
15	·555265	·481017	·417265	·362446	15
16	·533908	·458112	·393646	·338735	16
17	·513373	·436297	·371364	·316574	17
18	·493628	·415521	·350344	·295864	18
19	·474642	·395734	·330513	·276508	19
20	·456387	·376889	·311805	·258419	20
21	·438834	·358942	·294155	·241513	21
22	·421955	·341850	·277505	·225713	22
23	·405726	·325571	·261797	·210947	23
24	·390121	·310068	·246979	·197147	24
25	·375117	·295303	·232999	·184249	25
26	·360689	·281241	·219810	·172195	26
27	·346817	·267848	·207368	·160930	27
28	·333477	·255094	·195630	·150402	28
29	·320651	·242946	·184557	·140563	29
30	·308319	·231377	·174110	·131367	30
31	·296460	·220359	·164255	·122773	31
32	·285058	·209866	·154957	·114741	32
33	·274094	·199873	·146186	·107235	33
34	·263552	·190355	·137912	·100219	34
35	·253415	·181290	·130105	·093663	35
36	·243669	·172657	·122741	·087535	36
37	·234297	·164436	·115793	·081809	37
38	·225285	·156605	·109239	·076457	38
39	·216621	·149148	·103056	·071455	39
40	·208289	·142046	·097222	·066780	40
41	·200278	·135282	·091719	·062412	41
42	·192575	·128840	·086527	·058329	42
43	·185168	·122704	·081630	·054513	43
44	·178046	·116861	·077009	·050946	44
45	·171198	·111297	·072650	·047613	45
46	·164614	·105997	·068538	·044499	46
47	·158283	·100949	·064658	·041587	47
48	·152195	·096142	·060998	·038867	48
49	·146341	·091564	·057546	.036324	49
50	·140713	·087204	·054288	·033948	50

TABLE III. *Showing the Amount of £1 improved at Compound Interest for any Number of Years not exceeding 50.*

Yrs.	4 per Cent.	5 per Cent.	6 per Cent.	7 per Cent.	Yrs.
1	1·040000	1·050000	1·060000	1·07000	1
2	1·081600	1·102500	1·123600	1·14490	2
3	1·124864	1·157625	1·191016	1·22504	3
4	1·169858	1·215506	1·262476	1·31079	4
5	1·216652	1·276281	1·338225	1·40255	5
6	1·265319	1·340095	1·418519	1·50073	6
7	1·315931	1·407100	1·503630	1·60578	7
8	1·368569	1·477455	1·593848	1·71818	8
9	1·423311	1·551328	1·689478	1·83845	9
10	1·480244	1·628894	1·790847	1·96715	10
11	1·539454	1·710339	1·898298	2·10485	11
12	1·601032	1·795856	2·012196	2·25219	12
13	1·665073	1·885649	2·132928	2·40984	13
14	1·731676	1·979931	2·260903	2·57853	14
15	1·800943	2·078928	2·396558	2·75903	15
16	1·872981	2·182874	2·540351	2·95216	16
17	1·947900	2·292018	2·692772	3·15881	17
18	2·025816	2·406619	2·854339	3·37293	18
19	2·106849	2·526950	3·025599	3·61652	19
20	2·191123	2·653297	3·207135	3·86968	20
21	2·278768	2·785962	3·399563	4·14056	21
22	2·369918	2·925260	3·603537	4·43040	22
23	2·464715	3·071523	3·819749	4·74052	23
24	2·563304	3·225099	4·048934	5·07236	24
25	2·665836	3·386354	4·291870	5·42743	25
26	2·772469	3·555672	4·549382	5·80735	26
27	2·883368	3·733456	4·822345	6·21386	27
28	2·998703	3·920129	5·111686	6·64883	28
29	3·118651	4·116135	5·418387	7·11425	29
30	3·243397	4·321942	5·743491	7·61225	30
31	3·373133	4·538039	6·088100	8·14511	31
32	3·508058	4·764941	6·453386	8·71527	32
33	3·648381	5·003188	6·840589	9·32533	33
34	3·794316	5·253347	7·251025	9·97811	34
35	3·946088	5·516015	7·686086	10·6765	35
36	4·103932	5·791816	8·147252	11·4239	36
37	4·268089	6·081406	8·636087	12·2236	37
38	4·438813	6·385477	9·154252	13·0792	38
39	4·616365	6·704751	9·703507	13·9948	39
40	4·801020	7·039988	10·285717	14·9744	40
41	4·993061	7·391988	10·902861	16·022670	41
42	5·192784	7·761588	11·557033	17·144257	42
43	5·400495	8·149667	12·250455	18·344355	43
44	5·616519	8·557150	12·985482	19·628460	44
45	5·841176	8·985008	13·764611	21·002452	45
46	6·074823	9·434258	14·590487	22·472623	46
47	6·317816	9·905971	15·465917	24·045707	47
48	6·570528	10·401270	16·393872	25·728907	48
49	6·833349	10·921333	17·377504	27·529930	49
50	7·106683	11·467400	18·420154	29·457025	50

TABLE IV. *Showing the Present Value of £1 per Annum (Annuity) for any Number of Years not exceeding 50.*

Yrs.	4 per Cent.	5 per Cent.	6 per Cent.	7 per Cent.	Yrs.
1	1·000000	1·000000	1·000000	1·00000	1
2	2·040000	2·050000	2·060000	2·07000	2
3	3·121600	3·152500	3·183600	3·21490	3
4	4·246464	4·310125	4·374616	4·43994	4
5	5·416322	5·525631	5·637092	5·75073	5
6	6·632975	6·801912	6·975318	7·15329	6
7	7·898294	8·142008	8·393837	8·65402	7
8	9·214226	9·549108	9·897467	10·2598	8
9	10·582795	11·026564	11·491315	11·9779	9
10	12·006107	12·577892	13·180794	13·8164	10
11	13·486351	14·206787	14·791642	15·7836	11
12	15·025805	15·917126	16·869941	17·8884	12
13	16·626837	17·712982	18·882137	20·1406	13
14	18·291911	19·598631	21·015065	22·5504	14
15	20·023587	21·578563	23·275969	25·1290	15
16	21·824531	23·657491	25·672528	27·8880	16
17	23·697512	25·840366	28·212879	30·8402	17
18	25·645412	28·132384	30·905652	33·9990	18
19	27·671229	30·539003	33·759991	37·3789	19
20	29·778078	33·065954	36·785591	40·9954	20
21	31·969201	35·719251	39·992726	44·8651	21
22	34·247969	38·505214	43·392290	49·0057	22
23	36·617888	41·430475	46·995827	53·4361	23
24	39·082604	44·501998	50·815577	58·1766	24
25	41·645908	47·727098	54·864512	63·2490	25
26	44·311744	51·113453	59·156382	68·6764	26
27	47·084214	54·669126	63·705765	74·4838	27
28	49·967582	58·402582	68·528111	80·6976	28
29	52·966286	62·322711	73·639798	87·3465	29
30	56·084937	66·438847	79·058186	94·4607	30
31	59·328355	70·760789	84·801677	102·073	31
32	62·701468	75·298829	90·889778	110·218	32
33	66·209527	80·063770	97·343164	118·933	33
34	69·857908	85·066959	104·183754	128·258	34
35	73·652224	90·320307	111·434779	138·236	35
36	77·598313	95·836322	119·120866	148·913	36
37	81·702246	101·628138	127·268118	160·337	37
38	85·970336	107·708545	135·904205	172·561	38
39	90·409149	114·095023	145·058458	185·640	39
40	95·025515	120·799774	154·761965	199·635	40
41	99·826536	127·83976	165·09768	214·60957	41
42	104·81960	135·23175	175·95055	230·63224	42
43	110·01238	142·99334	187·50758	247·77650	43
44	115·41288	151·14301	199·75803	266·12085	44
45	121·02939	159·70016	212·74351	285·74931	45
46	126·87057	168·68516	226·50812	306·75176	46
47	132·94539	178·11942	241·09861	329·22439	47
48	139·26321	188·02539	256·56453	352·27009	48
49	145·83373	198·42666	272·95840	378·99900	49
50	152·66708	209·34800	290·33590	406·52893	50

TABLE V. *Showing the Amount of £1 per Annum (Annuity) forborne and improved at Compound Interest for any Number of Years not exceeding 50.*

Yrs.	4 per Cent.	5 per Cent.	6 per Cent.	7 per Cent.	Yrs.
1	0·961539	0·952381	0·943396	0·9345	1
2	1·886095	1·859410	1·833393	1·8080	2
3	2·775091	2·723248	2·673012	2·6243	3
4	3·629895	3·545951	3·465106	3·3872	4
5	4·451822	4·329477	4·212364	4·1001	5
6	5·242137	5·075692	4·917324	4·7665	6
7	6·002055	5·786373	5·582382	5·3892	7
8	6·732745	6·463213	6·209794	5·9712	8
9	7·435331	7·107822	6·801692	6·5152	9
10	8·110896	7·721735	7·360087	7·0235	10
11	8·760476	8·306414	7·886875	7·4986	11
12	9·385073	8·863252	8·383844	7·9426	12
13	9·985647	9·393573	8·852683	8·3576	13
14	10·563122	9·898641	9·294984	8·7454	14
15	11·118387	10·379658	9·712249	9·1079	15
16	11·652295	10·837770	10·105895	9·4466	16
17	12·165668	11·274066	10·477260	9·7632	17
18	12·659296	11·689587	10·827604	10·059	18
19	13·133939	12·085311	11·158117	10·335	19
20	13·590325	12·462210	11·469921	10·594	20
21	14·029159	12·821153·	11·764077	10·585	21
22	14·451114	13·163003	12·041582	11·061	22
23	14·856841	13·488574	12·303379	11·272	23
24	15·246962	13·798642	12·550358	11·469	24
25	15·622079	14·093945	12·783356	11·653	25
26	15·982768	14·375185	13·003166	11·825	26
27	16·329584	14·643034	13·210534	11·986	27
28	16·663062	14·898127	13·406164	12·137	28
29	16·983713	15·141074	13·590721	12·277	29
30	17·292032	15·372451	13·764831	12·409	30
31	17·588492	15·592810	13·929086·	12·531	31
32	17·873550	15·802677	14·084044	12·646	32
33	18·147644	16·002549	14·230230	12·753	33
34	18·411196	16·192904	14·368141	12·854	34
35	18·664612	16·374194	14·498247	12·947	35
36	18·908280	16·546852	14·620987	13·035	36
37	19·142577	16·711287	14·736780	13·117	37
38	19·367863	16·867893	14·846019	13·193	38
39	19·584483	17·017041	14·949075	13·264	39
40	19·792772	17·159086	15·046297	13·331	40
41	19·993052	17·294368	15·138016	13·394120	41
42	20·185627	17·423208	15·224543	13·452449	42
43	20·370795	17·545912	15·306173	13·506962	43
44	20·548841	17·662773	15·383182	13·557908	44
45	20·720040	17·774070	15·455832	13·605522	45
46	20·884654	17·880067	15·524370	13·650020	46
47	21·042936	17·981016	15·589028	13·691608	47
48	21·195131	18·077158	15·650027	13·730474	48
49	21·341472	18·168722	15·707572	13·766799	49
50	21·482185	18·255925	15·761861	13·800746	50

Q

Section II.

FOREIGN WEIGHTS AND MEASURES.

Comparison of French Weights and Measures with English.

Old System.*

The Paris pound = 7561 ⎫
    "   ounce = 472·5625 ⎪
    "   gros = 59·0703 ⎬ English troy grains.
    "   grain = 00·8204 ⎭

The toise = 6 French feet; the league = 2282 toises.

The Paris royal foot of 12 inches = 12·7977 ⎫
The inch   -   -   = 1·0659 ⎬ English inches.
The line, or one-twelfth of an inch = ·0074 ⎭

The Paris cubic foot = 1·211273 English cubic foot.
The cubic inch = 1·21063 English cubic inch.
The square inch = 1·13582 English square inch.
An arpent = $\frac{5}{8}$ English acre; arpent royal = $1\frac{1}{2}$ English acre.

Measures of Capacity. — The Paris pint contains 58·145 English cubical inches; and the English wine pint contains 28·875 cubical inches; or the Paris pint contains 2·0171082 English pints; therefore, to reduce the Paris pint to the English, multiply by 2·0171082.

The New System.

Lineal Measure.

	Eng. inches.			Eng. inches.
Millimètre =	·03937		Decamètre =	393·71
Centimètre =	·39371		Hecatomètre =	3937·1
Decimètre =	3·9371		Chiliomètre =	39371
Mètre †	= 39·371		Myriomètre	= 393710

	miles.	p.	yds.	ft.	in.
A decamètre =	0	0	10	2	9·7
A hecatomètre =	0	0	109	1	·1
A chiliomètre =	0	4	213	1	10·2
A myriomètre =	6	1	156	0	·5

8 chiliomètres are nearly = 5 English miles.

Note. — *Deca* prefixed, denotes 10 times; *hecato*, 100 times; *chilio*, 1000 times, &c. On the other hand *deci, centi, milli,* denotes the 10th, 100th, 1000th part, &c. So that *mètre* is the element of long measures; *are*, of superficial measures; *stere*, that of solid measures; and *litre* is the element of the measures of capacity; also, *gramme* is the element of all weights, being itself the weight of a cubic centimètre of distilled

---

* Still used in many parts.   † An English inch = ·0534 mètres.

water. The *are* is the square decamètre; the *litre,* the cubic deci-
mètre; the *stere,* the cubic mètre. The mètre itself is one ten-millionth
part of the terrestrial arc intercepted between the equator and the
north pole, as determined by the actual measurement of degrees in
different latitudes.

Measures of Capacity.

	Eng. cub. inch.			Eng. cub. inch.
Millilitre	= ·06102		Decalitre =	610·24429
Centilitre	= ·61024		Hecatolitre =	6102·44288
Decilitre	= 6·10244		Chiliolitre =	61024·42878
Litre	= 61·02442		Myriolitre =	610244·28778

A litre is nearly 2⅛ wine pints = 1·7607 English pint = ·22 gallons.
14 decilitres are nearly 3 wine pints.
A decalitre = 2·2009 Eng. gallons; a hecatolitre = 22·009 Eng. gal.
A chiliolitre is 1 tun, 12·75 wine gallons.

Measures of Weight.

	Eng. grains.			Eng. grains.
Milligramme =	·0154		Hecatogramme =	1544·4023
Centigramme =	·1544		Chiliogramme	
Decigramme =	1·5444		(Kilogram) } =	15444·0234
Gramme =	15·4440		Myriogramme	= 154440·2344
Decagramme =	154·4402			

A gramme = 15·444 gr. = 0·644 pennyweights = ·03216 oz. troy.
A decagramme is 6 dwts. 10·44 gr. troy; or 5·65 dr. avoirdupois.
A hecatogramme is 3 oz. 8·5 dr. avoirdupois.
A chiliogramme is 2 lb. 3 oz. 5 dr. avoirdupois = 2·68 lb. troy.
A myriogramme is 22 lbs. 1·15 oz. avoirdupois.
100 myriogrammes are 1 ton, wanting 32·8 lbs. or = 2207·2 lbs.

Agrarian Measure.

Mètre carré = 1·196033 square yards English.
Are, 1 square decamètre = 119·6046 English yards = 3·95 perches =
·0988 roods.
Decare = 1196·046 English yards = 39·5 perches.
Hecatare = 11960·46 English yards = 2 acres, 1 rood, 30·1 perches.

Solid Measure.

Decistre ⅒th stere	= 3·5315	cubic feet English, *for fire-wood.*
Stere, 1 cubic metre =	35·315	cubic feet English.
Decastere	= 353·15	cubic feet English.

Division of the Circle.

French.		English.	
100 seconds	= 1 minute.	60 seconds = 1 minute.	
100 minutes	= 1 degree.	60 minutes = 1 degree.	
100 degrees	= 1 quadrant.	360 degrees = 1 circle.	
4 quadrants	= 1 circle.		

*Proportion of Long Measures of different Nations to the English Foot,
which, for the sake of Comparison, is divided into* 1000 *Parts.*

English foot	- 1000	Rhinland	- 1033	Dantzic	- 944
Paris	- 1068	Strasburg	- 951	Danish -	- 1042
Venetian	- 1162	Nuremberg	- 1000	Swedish	- 977¾

Itinerary Measures of European Nations.

	Eng. miles.		Eng. miles.
French league is about	- 2¾	An Italian mile is about	- 1½
A German mile -	- 4	A Spanish league -	- 3⅜
A Dutch mile -	- 3¼	A Russian verst -	- ¾

SECTION III.

MATHEMATICAL TABLES.

Table of Square and Cube Roots.

No.	Sq. root.	Cube root.	No.	Sq. root.	Cube root.	No.	Sq. root.	Cube root.
1	1·	1·	340	18·4390	6·979	680	26·0768	8·793
10	3·1622	2·154	350	18·7082	7·047	690	26·2678	8·836
20	4·4721	2·714	360	18·9736	7·113	700	26·4575	8·879
30	5·4772	3·107	370	19·2353	7·179	710	26·6458	8·921
40	6·3245	3·419	380	19·4935	7·243	720	26·8328	8·962
50	7·0710	3·684	390	19·7484	7·306	730	27·0185	9·004
60	7·7459	3·914	400	20·	7·368	740	27·2029	9·045
70	8·3666	4·121	410	20·2484	7·428	750	27·3861	9·085
80	8·9442	4·308	420	20·4939	7·488	760	27·5680	9·125
90	9·4868	4·481	430	20·7364	7·547	770	27·7488	9·165
100	10·	4·641	440	20·9761	7·605	780	27·9284	9·205
110	10·4880	4·791	450	21·2132	7·663	790	28·1069	9·244
120	10·9544	4·932	460	21·4476	7·719	800	28·2842	9·283
130	11·4017	5·065	470	21·6794	7·774	810	28·4604	9·321
140	11·8321	5·192	480	21·9089	7·829	820	28·6356	9·359
150	12·2474	5·313	490	22·1359	7·883	830	28·8097	9·397
160	12·6491	5·428	500	22·3606	7·937	840	28·9827	9·435
170	13·0384	5·539	510	22·5831	7·989	850	29·1547	9·472
180	13·4164	5·646	520	22·8035	8·041	860	29·3257	9·509
190	13·7840	5·748	530	23·0217	8·092	870	29·4957	9·546
200	14·1421	5·848	540	23·2379	8·143	880	29·6647	9·582
210	14·4913	5·643	550	23·4520	8·193	890	29·8328	9·619
220	14·8323	6·036	560	23·6643	8·242	900	30·	9·654
230	15·1657	6·126	570	23·8746	8·291	910	30·1662	9·690
240	15·4919	6·214	580	24·0831	8·339	920	30·3315	9·725
250	15·8113	6·299	590	24·2899	8·387	930	30·4959	9·761
260	16·1245	6·382	600	24·4948	8·434	940	30·6594	9·795
270	16·4316	6·463	610	24·6981	8·480	950	30·8220	9·830
280	16·7332	6·542	620	24·8997	8·527	960	30·9838	9·864
290	17·0293	6·619	630	25·0998	8·572	970	31·1448	9·898
300	17·3205	6·694	640	25·2982	8·617	980	31·3049	9·932
310	17·6068	6·767	650	25·4950	8·662	990	31·4642	9·966
320	17·8885	6·839	660	25·6904	8·706	999	31·6069	9·996
330	18·1659	6·910	670	25·8843	8·750			

A Table of the Areas, Circumferences of Circles, and Sides of equal Squares corresponding to all Diameters, from 1 to 100.

Diameter.	Area.	Circumference.	Side of equal Square.
1·00	0·78539	3·141592	0·88622
2·	3·141592	6·283185	1·77245
3·	7·068583	9·424777	2·65868
4·	12·566370	12·566370	3·54490
5·	19·634954	15·707963	4·43113
6·	28·274333	18·849555	5·31736
7·	38·484560	21·991148	6·20358
8·	50·265482	25·132741	7·08981
9·	63·617251	28·274333	7·97604
10·	78·539816	31·415926	8·86226
11·	95·033177	34·557519	9·74849
12·	113·097335	37·699111	10·63472
13·	132·732289	40·840704	11·52095
14·	153·938040	43·982297	12·40717
15·	176·714586	47·123889	13·29340
16·	201·061929	50·265482	14·17963
17·	226·980069	53·407075	15·06585
18·	254·469004	56·548667	15·95208
19·	283·528736	59·690260	16·83831
20·	314·159265	62·831853	17·72453
21·	346·360590	65·973445	18·61076
22·	380·132711	69·115038	19·49699
23·	415·475628	72·256631	20·38321
24·	452·389342	75·398223	21·26944
25·	490·873852	78·539816	22·15567
26·	530·929158	81·681408	23·04190
27·	572·555261	84·823001	23·92812
28·	615·752160	87·964594	24·81435
29·	660·519855	91·106186	25·70058
30·	706·858347	94·247779	26·58680
31·	754·767635	97·389372	27·47303
32·	804·247719	100·530964	28·35926
33·	855·298599	103·672557	29·24548
34·	907·920276	106·814150	30·13171
35·	962·112750	109·955742	31·01794
36·	1017·876019	113·097335	31·90416
37·	1075·210085	116·238928	32·79039
38·	1134·114947	119·380520	33·67662
39·	1194·590606	122·522113	34·56285
40·	1256·637041	125·663706	35·44907
41·	1320·254312	128·805298	36·33530
42·	1385·442360	131·946891	37·22153
43·	1452·201204	135·083484	38·10775
44·	1520·530844	138·230076	38·99398
45·	1590·431280	141·371669	39·88021
46·	1661·902513	144·513262	40·76643
47·	1734·944542	147·654854	41·65266
48·	1809·557368	150·796447	42·53889
49·	1885·740990	153·938040	43·42511

Diameter.	Area.	Circumference.	Side of equal Square
50·	1963·495408	157·079632	44·31134
51·	2042·820622	160·221225	45·19757
52·	2123·716633	163·362817	46·08380
53·	2206·183440	166·504410	46·97002
54·	2290·221044	169·646003	47·85625
55·	2375·829444	172·787595	48·74248
56·	2463·008640	175·929188	49·62870
57·	2551·758632	179·070781	50·51493
58·	2642·079421	182·212373	51·40116
59·	2733·971006	185·353966	52·28738
60·	2827·433388	188·495559	53·17364
61·	2922·466566	191·637151	54·05984
62·	3019·070540	194·778744	54·94606
63·	3117·245310	197·920337	55·83229
64·	3216·990677	201·061929	56·71852
65·	3318·307240	204·203522	57·60475
66·	3421·194399	207·345115	58·49097
67·	3525·652355	210·486707	59·37720
68·	3631·681107	213·628300	60·26343
69·	3739·280655	216·769893	61·14965
70·	3848·451000	219·911485	62·03588
71·	3959·192141	223·053078	62·92211
72·	4071·504079	226·194671	63·80833
73·	4185·386812	229·336263	64·69456
74·	4300·840342	232·477856	65·58079
75·	4417·864669	235·619449	66·46701
76·	4536·459791	238·761041	67·35324
77·	4656·625710	241·902634	68·23947
78·	4778·362426	245·044226	69·12570
79·	4901·669937	248·185819	70·01192
80·	5026·548245	251·327412	70·89815
81·	5152·997350	254·469004	71·78438
82·	5281·017250	257·610597	72·67060
83·	5410·607947	260·752190	73·55683
84·	5541·769440	263·893782	74·44306
85·	5674·501730	267·035375	75·32928
86·	5808·804816	270·176968	76·21551
87·	5944·678698	273·318560	77·10174
88·	6082·123377	276·460153	77·98796
89·	6221·138852	279·601746	78·87419
90·	6361·725123	282·743338	79·76042
91·	6503·882191	285·884931	80·64669
92·	6647·610054	289·026524	81·53287
93·	6792·908715	292·168116	82·41910
94·	6939·778171	295·309709	83·30533
95·	7088·218424	298·451302	84·19155
96·	7238·229473	301·592894	85·07778
97·	7389·811319	304·734487	85·96406
98·	7542·963961	307·876080	86·85023
99·	7697·687399	311·017672	87·73646
100·	7853·981633	314·159265	88·62269

TABLE *showing the Power of various Species of Fuel, in reference to Hothouses, and other Buildings heated by Steam.*

Species of Fuel.	Effect in lb. of Water heated one Degree by one lb. of Fuel.	Effect in lb. of Water converted into Steam of 220°.	Quantity to convert a Cubic Foot of Water into low-pressure Steam.	Quantity to convert a Cubic Foot of Water into Steam, allowing 10 per cent. for loss.	Will melt of Ice.
Caking coal	9800 lb.	8·4 lb.	7·45 lb.	8·22 lb.	90 lb.
Coke	9000 —	7·7 —	8·1 —	9·00 —	94 —
Splint coal	7900 —	6·75 —	9·25 —	10·28 —	——
Oak wood, dry	6000 —	5·13 —	12·2 —	13·6 —	92 —
Ordinary oak	3600 —	3·07 —	20·31 —	22·6 —	——
Peat compact, of ordinary dryness	3250 —	2·8 —	22·5 —	25·0 —	19 —

SECTION IV.

OF THE THERMOMETER.

THE thermometer is an instrument for measuring the temperature of bodies, or the degree of intensity of their sensible heat. There are three different sorts of thermometers employed: thus, 1st, Fahrenheit's, used chiefly in Britain, Holland, and North America, the freezing point of which is at 32°, and the boiling point at 212°. 2nd. Reaumur's, used in France before the revolution, and now in Spain and elsewhere; the freezing point is *zero* (0), the boiling at 80°. 3rd, the Celsius, or Centigrade, now used everywhere by men of science, in which zero or freezing point is 0°, and the boiling point 100°.

We have noticed that the freezing point of Fahrenheit's thermometer is marked 32°, and the reason for this is said to have been, that this artist thought he had produced the greatest degree of cold possible, by a mixture of snow and salt, and the point at which the thermometer then stood in this temperature he marked *zero*. The point at which mercury begins to boil, he conceived to be the greatest degree of heat, and he made this the limit of his scale. He divided the distance between these two points into 600 equal parts, or degrees, and by trials he found that the mercury stood at 32°, or at the 32 division, *when water began to freeze;* and this station he called the *freezing point.* When the tube was put into boiling water the mercury rose to 212°, which is therefore the boiling point, and it is 180° above the freezing point; for $180° + 32° = 212°$.

In De l'Isle's thermometer, the whole bulk of the mercury when placed in boiling water is conceived to be divided into 100,000 equal parts, and from this one fixed point, the various degrees of heat, either above or below it, are marked in these parts on the scale by the various expansions or contractions of the mercury in all the imaginable varieties of heat.

In Reaumur's thermometers, or more properly De Luc's, the scale begins at the freezing point, which is marked 0 or *zero;* and the point to which the mercury rises when the thermometer is in boiling water is marked 80°, which of course corresponds with 212° of Fahrenheit's. The thermometer of Celsius has 100° between the freezing point and that of boiling water. The temperatures indicated by any of these thermometers may be reduced to the corresponding degrees of any of the others by the following rules : —

To reduce the degrees of temperature of the Centigrade thermometer, and of that of Reaumur, to degrees of Fahrenheit's scale, and conversely, we have —

Rule 1. Multiply the Centigrade degrees by 9 and divide the product by 5, or multiply the degrees of Reaumur by 9 and divide the product by 4 ; then add 32 to the quotient in either case, and the sum is the degrees of temperature on Fahrenheit's scale.

Rule 2. From the number of degrees on Fahrenheit's scale subtract 32; multiply the remainder by 5 for Centigrade degrees, or by 4 for those of Reaumur's scale, and the product in either case being divided by 9 will give the temperature required.

Note.—We have given this introduction and these rules in order that the reader may have a full knowledge of a useful instrument, without which we cannot regulate the temperature of hothouses accurately. If there should be one thermometer of each sort in a house, he would then see the accuracy of our table, and the truthfulness of the foregoing rules.

Fahr.	Reaum.	Cen.	Fahr.	Reaum.	Cen.	Fahr.	Reaum.	Cen.	Fahr.	Reaum.	Cen.
212	80	100	153	53.7	67.2	94	27.5	34.4	35	1.3	1.6
211	79.5	99.4	152	53.3	66.6	93	27.1	33.8	34	0.8	1.1
210	79.1	98.8	151	52.8	66.1	92	26.6	33.3	33	0.4	0.5
209	78.6	98.3	150	52.4	65.5	91	26.2	32.7	32	0	0
208	78.2	97.7	149	52	65	90	25.7	32.2	31	—0.4	—0.5
207	77.7	97.2	148	51.5	64.4	89	25.3	31.6	30	—0.8	—1.1
206	77.3	96.6	147	51.1	63.8	88	24.8	31.1	29	—1.3	—1.6
205	76.8	96.1	146	50.6	63.3	87	24.4	30.5	28	—1.7	—2.2
204	76.4	95.5	145	50.2	62.7	86	24	30	27	—2.2	—2.7
203	76	95	144	49.7	62.2	85	23.5	29.4	26	—2 6	—3.3
202	75.5	94.4	143	49.3	61.6	84	23.1	28.8	25	—3.1	—3.8
201	75.1	93.8	142	48.8	61.1	83	22.6	28.3	24	—3.5	—4.4
200	74.6	93.3	141	48.4	60.5	82	22.2	27.7	23	—4	—5
199	74.2	92.7	140	48	60	81	21.7	27.2	22	—4.4	—5.5
198	73.7	92.2	139	47.5	59.4	80	21.3	26.6	21	—4.8	—6.1
197	73.3	91.6	138	47.1	58.8	79	20.8	26.1	20	—5.3	—6.6
196	72.8	91.1	137	46.6	58.3	78	20.4	25.5	19	—5.7	—7.2
195	72.4	90.5	136	46.2	57.7	77	20	25	18	—6.2	—7.7
194	72	90	135	45.7	57.2	76	19.5	24.4	17	—6.6	—8.3
193	71.5	89.4	134	45.3	56.6	75	19.1	23.8	16	—7.1	—8.8
192	71.1	88.8	133	44.8	56.1	74	18.6	23.3	15	—7.5	—9.5
191	70.6	88.3	132	44.4	55.5	73	18.2	22.7	14	—8	—10
190	70.2	87.7	131	44	55	72	17.7	22.2	13	—8.4	—10.5
189	69.7	87.2	130	43.5	54.4	71	17.3	21.6	12	—8.8	—11.1
188	69.3	86.6	129	43.1	53.8	70	16.8	21.1	11	—9.3	—11.6
187	68.8	86.1	128	42.6	53.3	69	16.4	20.5	10	—9.7	—12.2
186	68.4	85.5	127	42.2	52.7	68	16	20	9	—10.2	—12.7
185	68	85	126	41.7	52.2	67	15.5	19.4	8	—10.6	—13.3
184	67.5	84.4	125	41.3	51.6	66	15.1	18.8	7	—11.1	—13.8
183	67.1	83.8	124	40.8	51.1	65	14.6	18.3	6	—11.5	—14.4
182	66.6	83.3	123	40.4	50.5	64	14.2	17.7	5	—12	—15
181	66.2	82.7	122	40	50	63	13.7	17.2	4	—12.4	—15.5
180	65.7	82.2	121	39.5	49.4	62	13.3	16.6	3	—12.8	—16.1
179	65.3	81.6	120	39.1	48.8	61	12.8	16.1	2	—13.3	—16.6
178	64.8	81.1	119	38.6	48.3	60	12.4	15.5	1	—13.7	—17.2
177	64.4	80.5	118	38.2	47.7	59	12	15	0	—14.2	—17.7
176	64	80	117	37.7	47.2	58	11.5	14.4	—1	—14.6	—18.3
175	63.5	79.4	116	37.3	46.6	57	11.1	13.8	—2	—15.1	—18.8
174	63.1	78.8	115	36.8	46.1	56	10.6	13.3	—3	—15.5	—19.4
173	62.6	78.3	114	36.4	45.5	55	10.2	12.7	—4	—16	—20.
172	62.2	77.7	113	36	45	54	9.7	12.2	—5	—16.4	—20.5
171	61.7	77.2	112	35.5	44.4	53	9.3	11.6	—6	—16.8	—21.1
170	61.3	76.6	111	35.1	43.8	52	8.8	11.1	—7	—17.3	—21.6
169	60.8	76.1	110	34.6	43.3	51	8.4	10.5	—8	—17.7	—22.2
168	60.4	75.5	109	34.2	42.7	50	8	10	—9	—18.2	—22.7
167	60	75	108	33.7	42.2	49	7.5	9.4	—10	—18.6	—23.3
166	59.5	74.4	107	33.3	41.6	48	7.1	8.8	—11	—19.1	—23.8
165	59.1	73.8	106	32.8	41.1	47	6.6	8.3	—12	—19.5	—24.4
164	58.6	73.3	105	32.4	40.5	46	6.2	7.7	—13	—20	—25
163	58.2	72.7	104	32	40	45	5.7	7.2	—14	—20.4	—25.5
162	57.7	72.2	103	31.5	39.4	44	5.3	6.6	—15	—20.8	—26.1
161	57.3	71.6	102	31.1	38.8	43	4.8	6.1	—16	—21.3	—26.6
160	56.8	71.1	101	30.6	38.3	42	4.4	5.5	—17	—21.7	—27.2
159	56.4	70.5	100	30.2	37.7	41	4	5	—18	—22.2	—27.7
158	56	70	99	29.7	37.2	40	3.5	4.4	—19	—22.6	—28.3
157	55.5	69.4	98	29.3	36.6	39	3.1	3.8	—20	—23.1	—28.8
156	55.1	68.8	97	28.8	36.1	38	2.6	3.3	—21	—23.5	—29.4
155	54.6	68.3	96	28.4	35.5	37.	2.2	2.7	—22	—24	—30
154	54.2	67.7	95	28	35	36	1.7	2.2	—23	—24.4	—30.5

SECTION V.

DIGGING, WELL - SINKING, &c.

24 cubic feet of sand, 17 ditto of clay, 18 ditto of earth, 13 ditto of chalk, equal 1 ton.

1 cubic yard of earth *before* digging will occupy about $1\frac{1}{2}$ cubic yard *when dug*, and it contains 21 striked bushels, which is considered a *single load*, and double these quantities a *double load*.

A load of mortar is 27 cubic feet, and requires for its preparation 9 bushels of lime and 1 cubic yard of sand. The mass will lessen one-third in bulk when made into mortar.

In facing wells, or building walls with freestone, which is largely used for this purpose in many parts of the country, all the stones ought to be laid on their natural beds, otherwise they will *flush* at the joints, and the laminæ speedily give way to the action of the air and water.

TABLE *showing the Quantity of Earth to be removed, the Number of Bricks, and Quantity of Water in Imperial Gallons, contained in circular Wells for each Foot in depth.*

Diameter in the clear.	½ Brick rim.			1 Brick rim.			Content in Imperial Gallon.
	Cubic Feet of digging.	Number of Bricks.		Cubic Feet of digging.	Number of Bricks.		
		Laid dry.	In Mortar.		Laid dry.	In Mortar.	
ft. in.	ft. in.			ft. in.			
1 0	2 4	28	23	4 9	70	58	4·90
1 3	3 1	33	27	5 9	80	66	7·65
1 6	4 0	38	31	7 1	90	74	11·02
1 9	4 9	43	35	8 3	102	82	15·00
2 0	5 9	48	41	9 6	112	92	19·58
2 3	7 1	53	44	11 0	122	100	24·78
2 6	8 3	58	48	12 6	132	108	30·59
3 0	11 0	68	57	15 9	154	126	44·05
3 6	14 2	79	65	19 6	174	142	60·96
4 0	17 7	89	73	23 8	194	154	78·31
4 6	21 6	100	82	28 3	214	176	99·11
5 0	26 0	110	90	33 2	234	192	122·36
5 6	30 7	120	98	38 5	254	209	148·06
6 0	35 8	130	107	44 2	276	226	176·20
6 6	41 3	140	115	50 3	296	242	206·80
7 0	47 2	150	123	56 7	316	260	239·83
7 6	53 5	160	131	63 6	336	276	275·32
8 0	60 1	170	140	70 9	358	292	313·25
8 6	67 2	180	148	78 5	378	308	353·63
9 0	74 7	191	156	86 6	398	326	396·46
10 0	90 8	212	174	103 9	438	360	489·46

235

INDEX.

A.

ACCOUNTS, mode of ascertaining the accuracy of, 77.
———, nominal, 76.; personal, 76.; real, 76.
Addition of Approximate Decimals, 20.
——— of Decimals, 17.
——— of Vulgar Fractions, 7.
Aerial perspective, 215.
Allotments of land, to set out, 165.
Altitude, an angle of, how taken with the quadrant, 125.
Angle of altitude and depression, mode of taking with the quadrant, 125.
——— of incidence, 217.; of reflection, 217.
Angles, horizontal, mode of taking, 129.
——— of depression, how taken, 124.
——— of elevation, how taken, 124.
——— which the hour lines form with the meridian on a horizontal dial, 131.
Angular bird's-eye Perspective, 215.
——— perspective, 215.
——— perspective, to draw a house in, 213.
Annuities, 37.
——— certain, 37.
———, present worth of, 39.
Annuity tables, 219.
Apothecaries' weight, 24.
Approximate Decimals, 18.
———, addition of, 20.
———, subtraction of, 20.
Architectural drawing, 191.
——— mouldings, 195.
Architecture, the five orders of, 194.
Areas of polygons whose sides are 1, table of, 124.
Arithmetic, 2.
——— of Terminal Decimals, 17.

Arithmetical calculations, data for, 22.
Atmospheric pressure, effect of, on running liquids, 152.
Avoirdupois weight, 25.

B.

Bailiffs, book-keeping for, 55.
Bark, expense of peeling, loading on waggons, &c., 65.
———, mode of conducting sales of, 65.
Bartering interests in property, 43.
Bearings, horizontal, mode of taking, 129.
Bench marks, 178.
Book-keeping, 49.
——— for bailiffs, 55.
——— for farmers, 70.
——— for foresters, 63.
——— for gardeners, 49.
——— for land-stewards, 55.
——— for nurserymen, 67.
Bought-book for nurserymen, 69.
Breast-wheel, 139.

C.

Cash-book for farmers, 74, 75.
——— for foresters, 64.
——— for gardeners, 49.
——— for land-stewards, 60, 61.
——— for nurserymen, 70.
Centre of gravity, 135.
Chain, description of the, 154.
Chain-pump, 153.
Chords, scale of, 122.
Ciphers in decimals, use of, 13.
Circle in perspective, to draw, 215.
———, mode of finding its diameter and circumference, 104.
———, to find its area, 104.
Circles, table of the areas and circumferences of, 229.

THE END.

LONDON :
Printed by A. SPOTTISWOODE,
New-Street-Square.

Printed in the United States
By Bookmasters